Frank Sauer

Microbicides in Coatings

VINCENTZ
Providing Professional Information

Cover: Tiberius Gracchus/fotolia.com

Frank Sauer
Microbicides in Coatings
Hanover: Vincentz Network, 2017
European Coatings Library
ISBN 3-86630-897-3
ISBN 978-3-86630-897-8

Vincentz Network, Plathnerstr. 4c, 30175 Hanover, Germany
T +49 511 9910-033, F +49 511 9910-029, books@european-coatings.com

Discover further books from European Coatings Library at:
www.european-coatings.com/shop

Layout: Vincentz Network, Plathnerstr. 4c, 30175 Hanover, Germany

European Coatings Library

Frank Sauer

Microbicides in Coatings

Preface

"Everyone is trying to accomplish something big, not realising that life is made up of little things."

Frank Clark (1860 – 1936)

Microbicides are substances that represent two sides of the same coin. On one hand, they help to control microorganisms that are responsible for the deterioration of materials and for causing commercial damage worth billions of euros a year. On the other, they are regarded with suspicion because their action can have side-effects on humans or on the environment or both.

Microorganisms have been part of our biosphere for billions of years, during which time they have been extremely successful due to their ability to adapt to the most challenging of conditions. Human life as we know it would not have been possible without the tireless assistance of these tiny organisms. Our intention must therefore not be to combat germs wherever they are encountered – microbicidal measures should only be taken in situations where germs cause harm to humans, be it out of medical need or the need to protect a material.

This book seeks to provide an overview of the different aspects of material protection, covering the spectrum from basic information about the universe of microorganisms, to the innate properties of microbicides, to the state of the art and finally to legislative aspects. The biggest challenge in this regard has been deciding which of the key issues to select from the vast wealth of information available, without straying off course. It therefore goes without saying that it has not been possible to cover every detail in depth, as that would have substantially exceeded the scope of the book. Wherever appropriate, references are provided so that the reader can conduct further research.

The book also seeks to familiarise laboratory assistants, technicians, graduates, engineers and chemists with the principles of material protection in the field of coatings. However, it should also prove rewarding to business people with a basic knowledge of chemistry and biology.

I would like to thank all those colleagues who provided information on selected topics and proffered their advice and made various suggestions and recommendations.

My very special thanks go to my wife and my daughter for their endless patience during the preparation of this book and for their forbearance when I was so often unavailable for leisure pursuits, especially at the final stage of writing. Without their support, this book would not have been finalised in time. I am also greatly indebted to them for giving a reader's perspective of the book.

Langenfeld, May 2017
Frank Sauer

For Marlis and Melina

Contents

1 Introduction to microbicides

Surfaces determine our daily lives in manifold ways. They define the borderline between an interior and exterior domain and they are essential for giving form to physical objects. Consequently, surfaces play a key role in our living environment.

Nature creates surfaces in a huge variety of ways, be it in the form of inorganic matter, such as rock, soil, sand, gas, and water, or in the form of organic matter, such as plants and living creatures. At all times, surfaces are subject to interactions, such as approximation, adhesion, transformation, penetration, diffusion, attack and – in the worst case – destruction.

In general, coatings are designed to build a specific, well-defined layer on top of surfaces. Such layers can confer tremendous functionality: they can have a signalling function (e.g. a traffic sign), a commercial purpose (e.g. an advertising hoarding), an infrastructural purpose (e.g. a pavement), a protective function (e.g. a thermal insulation system also known as ETICS/EIFS [1]) and finally a decorative function by means of which they create an attractive appearance or convey a philosophical message, as in the case of paintings and other artwork.

In the construction field, the protective function of coatings is very often combined with a decorative purpose, e.g. protection of walls against energy loss in conjunction with an external layer for creating an attractive façade that retains its appeal in the long term. The materials used for designing coatings also vary enormously and they are usually used in combi-

Figure 1.1: Examples of different kinds of coating materials

[1] ETICS: External Thermal Insulation Composite System; EIFS: Exterior Insulation and Finishing System

nation. An architectural paint can be a quite complex mixture of polymeric or pre-polymeric binder, organic/inorganic solid matter, such as pigments, fillers and diverse additives and solvents for keeping the particles in the liquid phase, which, in this specific case, is a prerequisite for applying the coating material by a simple technique, such as brushing, rolling or spraying. But coatings, as a generic term, are not limited to paints and plasters. They can also consist of completely different materials applied by numerous other techniques, such as metal coverings made of gold, copper, brass, chromium, lead, titanium, and platinum (e.g. for roofing churches or producing prostheses as well as implants in medicine), glass panels which are widely used for façades of skyscrapers, ceramic tiles, especially in wet areas, and a huge swathe of plastics – to mention just a few (see Figure 1.1).

Most coatings have one property in common. Sooner or later, they become susceptible to attack and destruction. This might come about as a result of natural climatic conditions or other environmental factors, man-made physical and/or chemical impact, or seemingly unremarkable species which are minute and work under-cover but which have been extremely successful and effective for billions of years: microorganisms.

Bacteria, yeasts, fungi, algae and lichens are remarkably adaptive to different environmental situations and can find their specific ecological niche even in the most inhospitable of conditions. Even materials such as concrete, plastics and the like which were long thought to withstand microbial degradation are not exempt from such attack (Figure 1.2)[1].

Even more diverse than the world of coating materials is the huge variety of microbial species found in nature. All these germs have a specific preference for certain living conditions, such as acidic/alkaline media, aerobic/anaerobic surroundings or shade/sunlight areas[2]. In addition, germs can interact with each other and particular species are even known to engage in a form of communication[3].

Whatever the specific living conditions are, microorganisms compete with each other for space which, depending on its location, provides water, food, heat, essential minerals or

Figure 1.2: Microbial infestation on plastic, concrete, stone and roof tiles

UV light for photosynthesis[4]. Even if we could create an almost totally sterile surface, e.g. by treatment with a strong disinfectant, that same surface is very likely to be quickly re-colonised once the microbicide has disappeared or lost its effectiveness[5]. The survival or elimination of any given species is determined by its toughness, a fundamental principle of nature that can also be readily observed in the animal kingdom[6].

Just as humans have conquered the planet on a macroscopic level, microbial species have conquered the microcosm. Their success is due to an ability to rapidly adapt to changing external conditions that is significantly supported by their very high reproduction rates[7]. Only those germs which are capable of quickly developing a biological response to environmental stress and challenges will survive. The new, successful genetic material is then readily duplicated and passed down to subsequent generations.

The conquest of surfaces by these germs reflects two sides of the same coin. On one side, they are absolutely essential for human life (e.g. digestion, metabolism, acid mantle)[8]. Without the help of microbial species, human life as we know it would be inconceivable.

On the other, germs are responsible for the deterioration of substrates (Figure 1.3) and consequently for causing economic damage or even harm to plants, animals and humans. Recent decades have seen a remarkable increase in the microbial infestation of façades, for instance. The environmental conditions which microorganisms need to flourish in have improved considerably due to continuing eutrophication of the atmosphere with organic pollutants and to dramatic changes in the global climate[9]. In Germany, alone, the annual economic damage arising from microbially induced defacement and bio-corrosive deterioration is valued at EUR 8 to 16 billion[10, 11].

When it comes to providing protection for coatings, it is essential to strike the right balance between controlling germs so as to avoid disease and economic damage on one hand and tolerating microbial life where it is necessary and useful on the other. Thus, whole-scale eradication of microorganisms per se is not the right objective at all. Germs should only be

Figure 1.3: Examples of deteriorating substrates

combated in areas where they **adversely** impinge on the human sphere. This calls for an intelligent strategy for developing antimicrobial products which act not only where they are truly needed but also in dosage levels which minimise the adverse impact on the environment and humans and which simultaneously control the target germs with high efficacy, thereby avoiding the development of resistance. This point is discussed further in Section 3.

1.1 Classification of microorganisms

The term 'microorganism' derives from the Greek **μικρός (mikros)** "small" and **όργανισμός (organismós)** "organism" and describes a microscopic living species which consists of either unicellular or multicellular structures[12, 13]. Microorganisms were first discovered in 1674 by van Leeuwenhoek who observed bacteria through a single-lens microscope of his own design[13].

The classification of microorganisms is quite a complex area and a full treatment would be beyond the scope of this book. Consequently, only an overview is provided in this section. For comprehensive details, the reader should refer to specialist literature, such as[14–24]. The currently accepted classification of life forms recognises three domains[25]:
- Archaea
- Bacteria
- Eukaryota

Archaea and bacteria are subsumed under the term 'prokaryotes' meaning a unicellular organism with relatively simple cell compartment structures[26]. In contrast, 'eukaryotes' have much more sophisticated cell structures enclosed within membranes; in particular, they possess a membrane-bound cell nucleus. The names 'prokaryote' and 'eukaryote' are derived from the Greek **πρό (pro)** "before" and **ευ (eu)** "well"/"true" and **κάρυον (karyon)** "nut"/"kernel"[27–29].

1.1.1 Archaea

Archaea were initially dedicated to the domain bacteria ('archaebacteria'). However, this classification is outdated because these microorganisms possess unique properties which distinguish them from both bacteria and eukaryotes[30]. They were originally regarded as extremophiles that live in harsh environments, such as volcanic hot springs, but have since been found in a wide range of other natural habitats, such as fens, soil and sea water[31]. Archaea in plankton are believed to be one of the most abundant groups of organisms on earth[32]

Figure 1.4: Halophiles in water bodies of very high salt concentration

but they have also been found in the human colon and the navel[31]. Finally, archaea play an important part in global environmental processes, such as the carbon and nitrogen cycles[32].

Thanks to state-of-the art techniques employed in molecular biology, especially the polymerase chain reaction (PCR), archaea are now known to be widely distributed in nature and common in all habitats on earth[33]. The ribosomal[2] genes from species found in diverse environmental surroundings have been analysed and numerous organisms which had not been cultured in the laboratory thus far have now been classified as belonging to the archaea domain[34, 35]. Interesting examples of archaea species living under extreme conditions are hyperthermophiles[36] and halophiles[37]. Hyperthermophiles can thrive in very hot environments, sometimes even at temperatures exceeding 100 °C and under high pressure, e.g. on the walls of hydrothermal vents in the vast depths of the sea. One of the toughest species here is Strain 121 which can double its population in 24 hours at temperatures of 121 °C under pressure in an autoclave[38]. Hyperthermophiles were first discovered by Brock[39, 40] in hot springs in Yellowstone National Park in 1965. Since then, more than 70 hyperthermophiles species have been discovered[41].

Halophiles are found particularly in water bodies that have very high salt concentrations, such as the Great Salt Lake in Utah and the Dead Sea[42]. A subset of this species has a red appearance due to the presence of carotenoid compounds (Figure 1.4).

There is no clear evidence that archaea are pathogenic or parasitic, but they are known to act as *'commensals'*, which are organisms that cohabit with other organisms by using the same nutrient base. One example is methanogens, which profit from a supply of food in the guts of humans and ruminants and which in turn support digestion due to their vast numbers[32].

1.1.2 Bacteria

Bacteria and archaea evolved from an ancient common ancestor[25]. The first forms of life on earth appeared approximately 4 billion years ago in the form of unicellular microorganisms. For about 3 billion years, bacteria and archaea dominated the terrestrial habitats on our planet, such as soils, fens, water bodies and other natural compartments[43, 44], but nowadays they can even be detected on radioactive waste[45] and in manned spacecraft[46].

The total number of bacteria on earth is unimaginably high. The following examples illustrate how, in terms of 'head count', bacteria cells are by far the predominant life form on our planet, vastly outnumbering the world's human population:

- Scientists estimate the totality of bacteria on earth to be roughly 5×10^{30}[47], i.e. a five followed by 30 zeros. The respective biomass is greater than that of all the terrestrial plants and animals taken together[48].
- One gramme of soil contains approximately 40 million bacterial cells, and even natural freshwater has been found to host a million bacterial cells per millilitre[49].
- Approximately 80 million bacteria are exchanged during every intense kiss. However, this should not discourage couples from keeping this tradition, as bacterial exchange stimulates the human immune system and boosts the body's defences[50].

Even in the most extreme habitats on earth, such as the Mariana Trench at a depth of 11 kilometres researchers hypothesised the existence of bacteria there[51, 52].

[2] Ribosome: macromolecular complexes made of proteins and ribonucleic acids (RNA)

Also, human body parts can be extensively colonised by bacteria, especially the gut flora and the skin. There are roughly 10 times as many bacterial colonists in human flora as there are cells in the human body[53]. But most of them are harmless and some are beneficial. Only a minority are pathogenic and capable of causing various diseases[49].

Research on bacterial species is by no means exhausted, as only a small proportion of them have been fully characterised. In addition, many bacteria species could not yet been cultured in the laboratory[54].

Bacteria come in all shapes and sizes. Their cell dimensions are in the single-digit micrometre range, and so the cells are normally invisible to the unaided eye[49].
Spherical (*'coccus'*, pl. *'cocci'*) or rod-shaped (*'bacillus'*, pl. *'bacilli'*) bacteria are the most common species, but there are also spiral-shaped (*'spirella'*), comma-shaped (*'vibrio'*) or tightly coiled-shaped (*'spirochaetes'*) representatives, beside others[49]. Cell shape is important because it can help a given species to find food or to escape predators[55, 56].

Bacterial cells often aggregate in a specific way, e.g. in pairs, chains, clusters, and filaments. The last of these is often enclosed by a sheath containing diverse individual cells[49].

One important property in the context of cell aggregation is the formation of biofilms, which often occur at boundary layers. Biofilms can be detected particularly in aqueous systems, either on the water surface or at the interface with a solid phase (Figure 1.5). Biofilms are typically slimy layers containing embedded microorganisms, such as bacteria, algae, and fungi.

They can be regarded as a primary life form because the oldest fossils found to date originated from microorganisms in biofilms that lived 3.2 billion years ago[57]. The biofilm is a proven life form in view of the fact that it is so widespread in nature. The vast majority of microorganisms live in biofilms[58].

Biofilms contain water by way of main component, plus microorganisms and their secretion products, so-called extracellular polymeric substances (EPS) such as polysaccharides, proteins, lipids and nucleic acids. In combination with water, these biopolymers can form hydrogels, giving rise to a slimy matrix of more or less stable shape[59, 60].

Figure 1.5: Biofilm in a toilet flush

Different microbial strains usually coexist within a biofilm, so that, for example, in the space of just a few hundred micrometres aerobic and anaerobic bacteria can occur. The biofilm matrix is often interspersed with channels, pores and voids which allow for mass transfer and water supply. Inside a biofilm, mass transport mainly takes place by diffusion[59, 60].

Furthermore, microorganisms possess an intercellular communication system known as *'quorum sensing'*[61, 62]. The bacteria use this system to coordinate a set of processes in biofilms by activating genetic programmes in the various cells. Surface contact with corresponding microbial cells causes some genes to be switched on and others to be switched off. Specific signalling molecules enable microbial species to transmit information, and gene transfer with neighbouring cells is also known. Thus, a biofilm constitutes a flexible and capable life form that is essentially comparable to a more sophisticated, multicellular organism[57].

A biofilm provides effective shelter even in areas where maximum cleanliness is assumed (Figure 1.6). It protects against environmental stress, such as major fluctuations in pH and temperature, lack of food, and UV light. Finally, it confers a certain amount of protection against those chemical molecules which are designed to control germs: microbicides. This last property is possible because the chemical molecules either are unable to readily penetrate the biofilm or because unfavourable conditions in the biofilm hamper the efficacy of the microbicide.

It is known, for instance, that bacteria in a biofilm can retard their metabolism to such an extent that they enter a state of dormancy or stand-by mode. In this condition, they ingest virtually no antibiotic substances, thereby essentially protecting themselves against external stress[63]. Although this strategy of inactivity is successful at this micro level, it may not necessarily serve as a role model for the macrocosm.

Biofilms are ubiquitous. To mention only a few areas, they can be detected in soil, sediment, glacial ice, on rocks, plants and animals (here especially on mucous membranes), in technical equipment, such as pipes and tanks, and even in spacecraft[57, 64].

Human oral, dermal and intestinal florae are prominent examples of substantial bacterial communities which exist, as it were, right under our noses. A specific interaction occurs between the host and the bacteria involved. Germs which benefit from the host without impairing it are called *'commensals'*. When both the host and the bacteria benefit from each other, this interaction is called *'mutualism'*. Bacteria perform numerous tasks in such relationships. They are crucial for the maturation of our immune systems in the first years of our lives. They also prevent our body from being colonised by pathogenic germs and they support digestion processes. Any imbalance in this relationship can cause diseases[57].

Figure 1.6: Biofilm in a washing machine

Compared with this positive role in the aforementioned examples, biofilm formation in industrial production can have a severe negative impact on technical equipment. Parts of a given biofilm might be released from a boundary layer to the water phase (Figure 1.7) and translocated by the water flow to other macroscopic areas, thereby disseminating microbial life to other production areas, e.g. piping, filter equipment, valves, and tanks. Appropriate plant hygiene measures are essential for microbial control in plant components. In the worst-case scenario, production has to be halted completely in order that massive product contamination or a health threat posed to production staff (e.g. by *Legionella spp.*) may be dealt with.

A consequence of biofilm formation and unhindered proliferation is that the products of microbial degradation, such as acids, can trigger bio-corrosion that will gradually destroy the colonised substrates. Even alloyed steel is not immune to such attack. Just as dental enamel is attacked by plaque, with the formation of caries, contaminated materials in industrial production are likely to experience bio-corrosive attack as well. Microbial induced corrosion (MIC) can reduce the service life of a production site. In conjunction with potential shutdown time and increased energy consumption, this phenomenon is estimated to be responsible for financial losses in the double-digit billion euros range[65].

Biofouling is a preliminary stage of bio-corrosion. As the biofilm proliferates, macroscopic effects on substrates, such as discolouration, clogging, slime formation, gas evolution and release of odours, can be detected. At this stage, although the biofouling impairs performance, it does not damage the material[65].

Biofilms on underwater bodies, e.g. large container vessels and on maritime sensor systems, fall into this category. Biofouling on container ships is responsible for greatly increasing drag, leading to a substantial loss of vessel speed and higher fuel consumption. To solve this problem, industry has developed special antifouling paints for marine application that defer biofilm formation by releasing biocidal substances.

The surface of any given bacteria species is usually composed of a cell wall and a cell membrane, which together are called the *'cell envelope'*. The cell wall commonly comprises polysaccharide chains that are crosslinked by peptide molecules to form a large three-dimensional, net-like macromolecule known as a *'peptidoglycan'*[66]. This boundary layer confers stability and shape and is essential for keeping the bacterial cell alive. The cell walls of bacteria differ from those of archaea, which do not contain peptidoglycan. They are also distinct from the cell walls of fungi, which contain chitin instead of peptidoglycan[67].

As far back as 1884, Gram[68] developed a stain test that allows for a coarse classification of bacteria. This test is based upon the differential staining of cell walls as a function of structure (Figure 1.8). The first step consists is staining the bacteria under test with a solution of crystal violet (also known as pyoctanine blue) and phenol, followed by treatment with an iodine/iodide complex (Lugol's solution), which produces a deep blue appearance on all bacteria. In the second step, the coloured bacteria are treated (washed) with 96 percent alcohol. Bacterial cell walls of higher thickness remain deep blue (*'gram-positive'*) whereas thin cell wall structures are decolourised as the gram stain is washed off (*'gram-negative'*). When counterstained with a second dye (fuchsin or safranin), gram-negative bacteria appear red or red orange[68].

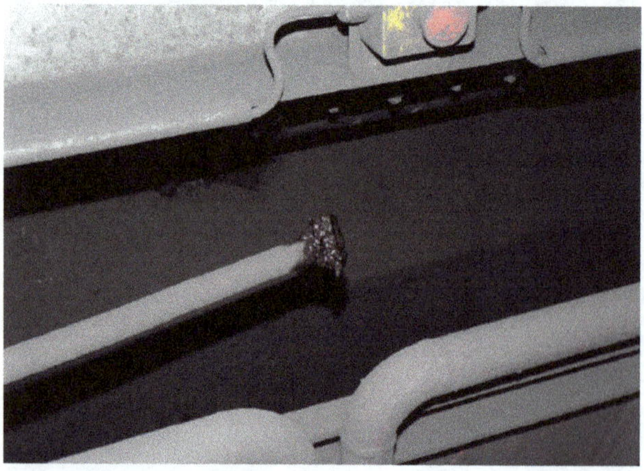

Figure 1.7: Parts of a biofilm disseminating microbial life to other production areas

Although this principle was established more than 130 years ago, it still represents state-of-the-art microbiological testing, because Gram staining is an important distinguishing feature and serves as a taxonomic criterion. Together with bacterial morphology, i.e. the study of bacterial shape, it enables many bacterial species to be allocated to one of the following groups:

- Gram-positive bacilli
- Gram-negative bacilli
- Gram-positive cocci
- Gram-negative cocci

Gram-positive bacteria include the genera *Streptococcus, Enterococcus, Staphylococcus, Listeria, Clostridium,* while *Escherichia, Salmonella, Klebsiella, Proteus, Enterobacter, Pseudomonas* **and** *Legionella* are gram-negative[69].

Not all bacteria can be classified by this technique, however, as there are also species that react in a variable or indefinite way to the Gram stain. For instance, tuberculosis-inducing Mycobacteria, such as *Mycobacterium tuberculosis,* contain within their cell walls mycolic acids, which are lipid substances that resist Gram staining[70]. The remedy in this case is to employ the *'Ziehl-Neelsen stain'*, also known as the *'acid-fast stain'*[71].

Many crucial biochemical processes take place across the interface between the cell interior and the exterior domain. As a consequence, the cell wall is a major point of attack for microbicides. One way to control germs, for instance, is to deploy antibiotics to inhibit peptidoglycan synthesis inside the bacterial cell[67].

The structural differences in cell walls, as described above, are also crucial to germ control, as some antibiotics are effective only against gram-positive bacteria, and not against gram-negative species[72].

A subset of gram-positive bacteria has the ability to form dormant structures without a detectable metabolism. Such *'endospores'*[73] can survive extreme conditions, e.g. in the space vacuum[74], and remain viable for millions of years[75, 76].

1.1.3 Eukaryotes

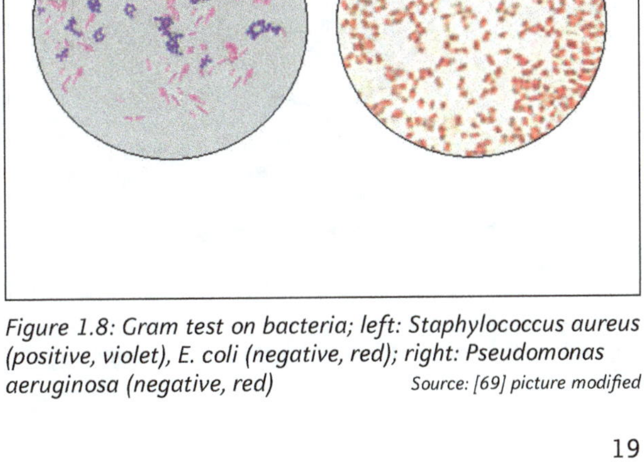

Figure 1.8: Gram test on bacteria; left: Staphylococcus aureus (positive, violet), E. coli (negative, red); right: Pseudomonas aeruginosa (negative, red) *Source: [69] picture modified*

The characteristic feature of eukaryotes (Figure 1.9) is the cell nucleus: this contains the genetic information (DNA) and is enclosed by a so-called nuclear envelope, a membrane consisting of two lipid bi-layers. The multiple pores in the membranes regulate the transport processes of diverse molecules, such as proteins, ribonucleic acids (RNA) which play an important role in biological processes, adenosine triphosphate (ATP) which is the universal and directly available energy carrier in cells, water, ions and other small molecules[77–79]. All multi-

Table 1.1: Contrasting properties of eukaryotes and prokaryotes

	Eukaryotes	**Prokaryotes**	
Domain	Eukaryota	Bacteria	Archaea
Cell nucleus	yes	no	no
Membrane-bound organelles	yes	no	no
Cell type	usually multicellular	usually unicellular	usually unicellular
Location of the genetic material	cell nucleus	cytoplasm	cytoplasm
Cell compartment structures	intense	poor	poor
Cell dimension [μm]	10–100	1–10	1–10
Examples	humans, animals, fungi, algae, plants & diverse unicellular species	E. coli	*Methanogenium sp.*

cellular organisms, including humans, animals, plants and fungi belong to the domain Eukarya. Moreover, eukaryotes contain further membrane-bound structures, called organelles. Examples here are the mitochondrion, which is the 'power plant of the cell,' and the Golgi apparatus which plays an important role in cell metabolism[80, 81]. Plants and algae additionally contain chloroplasts (Greek: **χλωρός (chlōrós)** "green" and **πλάστης (plastes)** "the one who forms")[82], organelles which can carry out photosynthesis[83].

Fungi

Fungi are eukaryotic microorganisms which have cells containing a mitochondrion and a cytoskeleton (Greek: **kýtos** "cell"), a network made of proteins featuring thin and filamentary cell sub-structures which give the cell shape and mechanical stability and which are responsible for movement and transports within the cell[84].

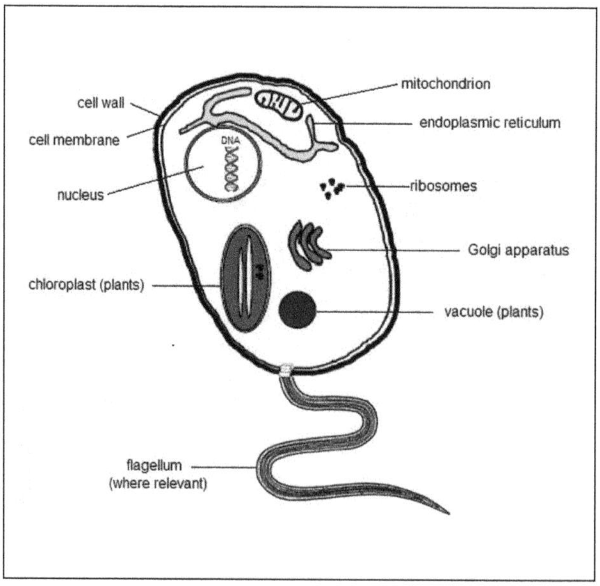

Figure 1.9: Illustration of important structural elements of eukaryotes

In the past, fungi were classified as belonging to the plant kingdom because of their immobile lifestyle. Nowadays, they are considered an independent kingdom by virtue of their physiological and genetic characteristics and would appear to be more related to the animal kingdom. One commonality between fungi and animals, for example, is the fact that both use glycogen for storing carbohydrates – plants employ starch for this purpose[85]. Most fungal cell walls contain chitin[86], unlike plant cell walls, which are usually made of cellulose[87].

What particularly distinguishes fungi from plants is a lack

of chlorophyll, which is essential for photosynthesis. Together with animals and the majority of bacteria and archaea, fungi are classified as being *'heterotrophic'*, i.e. they feed on organic matter in their surroundings by chemically degrading the organic material to use as an energy source and recycling its nutrients[88].

Fungi are important decomposing organisms (*'destruents'*) which, besides bacteria, are essential for the degradation of organic matter (Figure 1.10), such as cadavers, excrement, necrotic tissues and the like. By this method, large amounts of organic waste are continually returned to the eternal cycle of life[89]. Without, such an extremely effective recycling system, the human race would probably have 'choked' in debris for ages.

Fungi also play a very important role in the degradation and utilisation of biomolecules, such as lignin (an integral component wooden structures), cellulose, hemicellulose (an umbrella term for mixtures of polysaccharides of variable composition) and keratin (a major constituent of nails, claws, hooves, horn, barbs, feathers etc.). Lignin makes up from 20 % to 30 % of the dry weight of a wooden structure. Besides cellulose and chitin, this class of chemical ranks among the most common organic matter on earth[90]. By dint of teamwork among fungi, bacteria and animalcules, these biomolecules are eventually transformed into humus, which serves as an essential nutrient basis for the plant kingdom. In other words, microbial degradation causes organic matter to rise from the dead to generate new life, in the manner of the phoenix.

In contrast, many species of fungi benefit not just from dead matter but also from living plants; this poses a substantial threat – in the form of plant fungal diseases – to commercially important agricultural crops[91]. Without appropriate preventive or counter-measures, these diseases may lead to crop failure or even to a total loss of harvest. Many tree fungi also belong to this group of plant pathogens[92].

Fungal diseases do not just attack plants. Humans can be attacked by certain fungi species as well, especially the skin, hair, nails and mucous membranes. A multiplicity of bacteria and fungi typically colonise the human skin but do not cause any harm in normal circumstances. They settle in the upper dermal layers and feed on dead skin cells and sweat. However, factors such as stress, a compromised immune system and hormonal changes can trigger human diseases by fungal species that are usually harmless. That said, the benefits of fungal activities to humankind compensate for the disadvantages. For instance, fungi are employed in the production of alcohol, citric acid, and vitamin C and are even used for medical purposes. Production of the antibiotic penicillin is a prominent example of this. Certain species of fungi are important for the matu-

Figure 1.10: Degradation of organic matter by fungi

ration of dairy products, such as cheese. Brewer's yeast, wine lees and baker's yeast are well-known examples of beneficial microorganisms drawn from the group of unicellular fungi[88].

Fungi can be roughly classified by their propagation mode as either unicellular (e.g. yeasts) or multicellular organisms, so-called *'hyphae'*-forming fungi. Hyphae are cylindrical, threadlike structures several micrometres in diameter and up to several centimetres in length (Figure 1.11). Most fungi use hyphae for vegetative growth[93]. Propagation usually occurs at the hyphal tips where branching occasionally occurs, leading to a complex interconnected network that is collectively called a mycelium[94].

Hyphae are usually subdivided into cells which are separated from each other by a dividing wall called the septum. The latter contains pores that support an exchange of cytoplasm. In addition to chitin, the hyphal cell walls contain hemicellulose, lipids, proteins and other chemical compounds. The shape of hyphae may differ substantially from one fungal species to another. For instance, parasitic fungi often develop suckers that spread over vegetable cells in order to collect plant nutrients[88].

Most people have preconceptions of what fungi look like. But this biological structure with its characteristic shape, the so-called fruiting body (*'sporocarp'*), does not constitute the entire fungus. It is only a minor part of the entire organism which is responsible for reproduction, survival and dissemination via formation of spores. In fact, it is the capillary and in most cases the invisible network of hyphae (e.g. in soil or on wood) that constitutes the major part of a fungus[88].

In terms of size, fungi range from microscopic small species to readily identifiable, large mushrooms. The largest known fungus in the world belong to the genus *Armillaria* (honey fungus) and was found in a National Forest in Oregon, USA[95]. With a surface area of more than 8.8 km^2, an approximate age of 2400 years and an estimated weight of 600 tonnes, it is regarded as being one of the largest and oldest living organisms on earth. Fungi are thought to have been in existence for nearly 1 billion years[88]. Today, approximately 100,000 species of fungus are known, with some experts estimating their total number in the range of several millions[96].

Figure 1.11: Hyphae structures of fungi

Fungi usually thrive when the

- humidity is higher than 70 %
- environment is slightly acidic (pH: 4.6 to 6.5)
- temperature is in the range 10 to 35 °C

Species of black mould can often be detected on building façades. These fungi contain dark pigments (melanin) in the hyphae and in the spores that protect them from UV radiation[97]. As a consequence, the appearance of a façade can turn – in the worst case – from a primarily pale, pleasant colour to a black-spotted, ugly surface in the course of just a few years. In a way, this process bears comparison with human skin, which also contains melanin compounds. Sunbathing turns the skin colour from beige to brown (or red, given the wrong attitude). Excessive exposure to sunlight can eventually compromise the appearance of human skin, even to the point of causing pathological changes[98].

Some examples of the black mould ('Dermatiaceae') mentioned above are species such as *Ulocladium, Cladosporium, Alternaria, Aureobasidium, Stachybotrys,* and *Phoma. Alternaria* is usually one of the prime suspects when black mould occurs on exterior walls or façades.

Apart from impairing the appearance, surface-colonising fungi can impair the physical properties of the substrate, e.g. its wettability, drying of the material, and can even destroy mineral and metallic construction materials (bio-corrosion). Moreover, they can be responsible for the microbial degradation of organic matter by means of enzymatic attack. This causes metabolism of hydrocarbons and other chemical compounds, which are often converted into volatile products as a result[99].

Consequently, a fairly recent method for revealing evidence of mould infestation is based upon the analysis of organic metabolites generated by fungi, known as 'microbially volatile organic compounds' (MVOCs). In many cases, these MVOCs are fungi-specific and are fairly similar in molecular weight and chemical structure to the volatile organic compounds (VOCs) released from construction materials and home furnishings.

Release of MVOCs is of particular relevance indoors, although concentrations there are usually much lower (several nanogrammes per cubic meter) than in the case of VOCs[99]. Yet, because MVOCs have extremely low odour thresholds, residents can sense even very low concentrations of them. For instance, substances like geosmin ((4S,4aS,8aR)-4,8a-dimethyl-1,2,3,4,5,6,7,8-octahydronaphthalen-4a-ol, CAS No.19700-21-1) (Figure 1.12) have a distinct, earthy odour in ambient air[100].

Although further substances such as lower alcohols, aldehydes, ketones, aromatics, and terpenes are also typical fungal metabolites, they are unsuitable for the above-mentioned method, because their origins cannot always necessarily be attributed to fungal action, as they are also released by construction materials and others.

Besides emitting an odour, fungi can also severely impair human health[101]:

- Allergenic effects due to incorporation of fungal spores (the dose-response relationship is very complex because it depends on individual susceptibility and on the allergenic potential of the fungal spores). Modern detection methods have shown that approx. 5 % of the German population has become sensitised to fungal spores – and the trend is rising.
- Toxicity (some metabolites generated by certain fungal species known as mycotoxins, and some fungal cell-wall constituents (glucan compounds) can be toxic to humans. In addition, the fungal-induced release of interleukins that function as immune-system regulators in the skin and mucous membranes can be seen as an immunotoxic effect. However, it should be mentioned that such toxic effects due to indoor fungal contamination are quite unlikely).
- Infections (these mostly affect immune-compromised individuals and, like the toxic effects, are not usually triggered by indoor mould fungi)

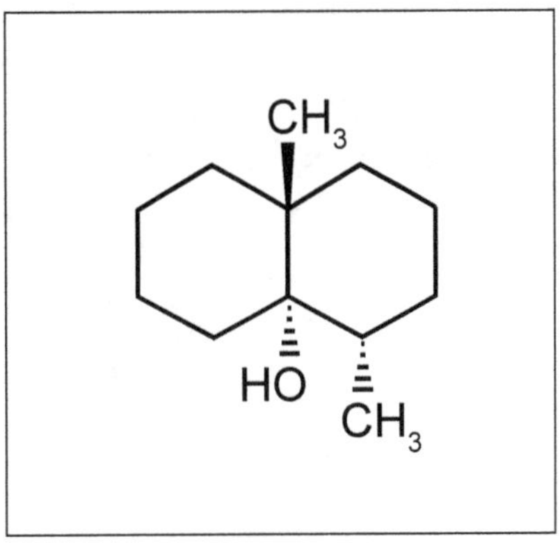

Figure 1.12: Molecular structure of geosmin

Algae

The name *'alga'* (pl. *algae*) comes from the Latin and means *"seaweed"*[102]. It does not describe a systematic group, as it is used more commonly as a generic term for certain life-forms that have some common traits but not a common ancestor[103]. For this reason, various algal species are not necessarily closely related. In particular, the term includes diverse eukaryotes ranging from unicellular and small-sized organisms (e.g. *Chlorella*, diatoms) to large multicellular representatives, such as giant kelp, which can be up to 50 metres long[104].

In the past, a group of prokaryotes was referred to in literature as *"blue-green algae"* because their appearance is similar to that of conventional algal species and because of their ability to engage in oxygen-emitting photosynthesis. However, it has emerged that these organisms must in fact be classified as bacteria[105] and so a better name for them is *'cyanobacteria'* which is derived from the Greek: **κυανός (kyanós)** "blue". It is a life-form that belongs to the oldest on our planet[106]. Unlike algae, cyanobacteria do not contain a real cell nucleus and so are not eukaryotes.

A more recent attempt at a definition is that algae are eukaryotic organisms which contain chlorophyll as primary pigment that supports photosynthesis but which are lacking coverings around their reproductive cells[107]. Most algal species are missing many of the distinct cell and tissue types found in terrestrial plants.

Algae are *'photoautotrophic'* organisms, i.e. they can produce complex organic molecules (carbohydrates, proteins etc.) from basic chemical compounds in their environment, such as carbon dioxide, water and minerals, by means of photosynthesis only. Like plants and cyanobacteria, they are producers in the food chain[103, 104] as opposed to *'heterotrophic'* species (Figure 1.13), which act as consumers (animals, fungi, and many bacteria and archaea). The generation of organic matter serves as an important nutrient source for aquatic life, but humans also benefit from algal action, e.g. in the form of the production of alginic acid, iodine, jelly (agar-agar, carrageen), organic fertilisers, feedstuffs and as raw material for the production of vegetable oil[108].

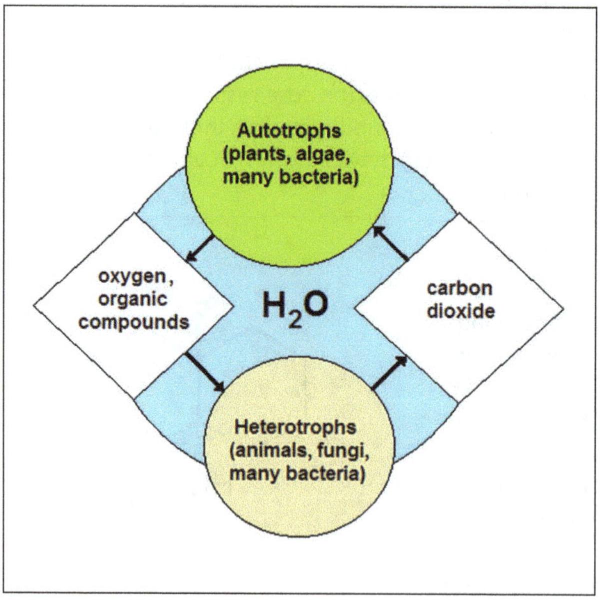

Figure 1.13: Interaction of heterotrophic and autotrophic species

Four prime examples taken from the most important groups of algae are highlighted below, along with selected properties[102, 109]:

- Diatoms (*Bacillariophyceae*): approx. 20,000 species. Predominantly found in marine compartments. As a special feature, their cell walls consist of silica. Although miniscule in size, they are so numerous that they produce the bulk of the oxygen in the atmosphere by assimilating carbon dioxide and sunlight. In addition, diatoms represent approx. 40 % of the biomass in the oceans[110].

- Green algae (*Chlorophyta*): more than 8,000 species, of which approx. 2/5ths are found in sea water and approx. 3/5ths in fresh water and rural areas.
- Red algae (*Rhodophyta*): more than 5,000 species, most prevalent in seaside regions, but also detectable in fresh water. Their photosynthetic active surface contains, in addition to the relevant type of chlorophyll and carotenoids, so-called accessory pigments (phycocyanine and phycoerythrine) which support photosynthesis by absorbing sunlight in a specific spectral range, i.e. light wavelengths that cannot be absorbed by chlorophyll itself. The corresponding energy is then transmitted to the chlorophyll molecule and utilised in the photosynthesis process. These accessory pigments out-trump the green colour of the chlorophyll and, as a consequence, give red algae their specific colour (red to purple).
- Brown algae (*Phaeophyceae*): approx. 1,500 species, almost entirely marine, range from small and delicate to large, extraordinarily rugged organisms.

Although algae are generally important organisms in water bodies, their occurrence is not limited to these areas. They are also detected on rocks, on trees, in desert soil, on glacial ice and last but not least on construction surfaces, such as façades, walls and damp glass panes.

Algae and cyanobacteria are often referred to in the literature as being the primary colonisers on surfaces, but it is not known exactly which of them appears first. While some authors suggest that algae, together with cyanobacteria, are the first colonisers, preparing the ground for heterotrophic organisms[111–114], such as fungi, others claim that cyanobacteria constitute the majority of the first organisms, conditioning the surfaces for microbial covering (with green algae, lichens and mosses)[115, 116, 119, 120]. In this connection, cyanobacteria are said to be typical *'ubiquitists'*, widely spread across all of the world's climate zones. They have even been found on Precambrian calcareous deposits dating back several hundred million to billions of years ago[115]. Whether green algae or cyanobacteria predominate seems to depend

Figure 1.14: Trentepohlia spp. featuring a strong reddish, orange or yellowish appearance

Figure 1.15: Lichens growing on a stone substrate

heavily on the ecological conditions, especially solar irradiation and substrate wettability[113]. In a study of painted buildings in Latin America, green algae were found to dominate the primary colonisation process[116]. This is supported by other research in Europe[117, 118] into the anti-algal activity of microbicides. Cyanobacteria were mainly isolated on untouched stone substrates[116]. Diatoms are said to be less prevalent on monuments in the European climate. They are especially found on works involving direct water contact or in littoral zones, while red algae have only been detected in rare cases in underground areas of high humidity or on constructions in very humid environments, e.g. memorial fountains. The correlation between microbial colonisation and the chemistry of the substrate is complex and has not yet been fully elucidated. It seems that some cyanobacteria species favour artificial substrates while others prefer the calcareous and siliceous types[116].

Besides the impact on aesthetics, the damage associated with this microbial action can be mainly assigned to the following processes[116]:

- Water retention and subsequent deterioration caused by freeze-thaw cycles
- Facilitated colonisation by other organisms (fungi, macroorganisms)
- Mobilisation of calcium ions, with subsequent detachment or dissolution of substrate areas (e.g. via production of lactic acid by *Trentepohlia aurea*)

Specialist literature[121] has described *Trentepohlia spp.* as being widespread, especially in humid climates (e.g. Singapore), where they establish large populations on construction buildings, as evidenced by their characteristic staining. These green algae contain large amounts of carotenoid pigments that mask the green chlorophyll and impart their strong reddish, orange or yellowish appearance (Figure 1.14).

Colonisation by algae and cyanobacteria in the human sphere is not limited to buildings and monuments. Previous research[122] has shown that algae have also been detected as the primary colonisers in technical installations, such as cooling towers. This can reduce the cooling capacity by as much as 20 %. Dead algal cells can additionally infiltrate water circulation systems, where their decomposition products provide nutrients for bacteria and fungi.

Lichens

The term *'lichen'* denotes a symbiotic community between a certain fungus and one or more partners capable of photosynthesis, i.e. algae and cyanobacteria[123]. The properties of lichens can differ substantially from those of their constituents. The typical habit of lichens is the outcome of symbiosis. The number of species worldwide is estimated at approx. 25,000, of which some 2,000 are found in central Europe. Lichens are named after the fungus partner in the symbiosis, as it is the one that provides the shape and structure. For this reason, lichens are classified as belonging to the kingdom of fungi, where they occupy an exceptional position as a distinct life-form. Thus, they do not belong to the plant kingdom[124].

Lichens are often found on tree bark in a broad diversity of colours, ranging from white to bright yellow, various shades of brown, strong orange, bright red, pink, olive green, blue-green, grey to deep black (Figure 1.15). Their food supply is either airborne, or obtained from precipitation water or from mineral intake.

Lichens are environmental indicators, especially for air quality, because the symbiotic relationship of the fungus and algal species is highly sensitive. Lichens absorb airborne nutrients and environmental pollutants nearly unfiltered as they do not possess roots for absorbing water from soil. They can therefore be severely damaged by pollution.

1.2 Microbicides

The term 'microbicide' is derived from the Greek: **mikros** "small", **bios** "life" and the Latin: **cadere** "to kill"[125]. It is used in this book as a generic term for substances that kill microorganisms. In particular, it covers antimicrobial compounds, such as bactericides, fungicides and algicides. However, the similar-sounding term 'biocide' is used here (especially in Section 5) in an even broader sense and encompasses not only microbicides but also other categories, including insecticides, herbicides and rodenticides. The purpose of this distinction is to highlight the fact that microbicides are not intended to attack the entire biosphere. Germs should be combated only in areas where they exert an **adverse** effect on humans or the environment and in accordance with the slogan: "As little as possible and as much as necessary".

Regulation (EU) No. 528/2012 of the European Parliament and of the Council of 22 May 2012 concerning the making available on the market and use of biocidal products[126] – also called the 'Biocidal Products Regulation' (BPR) – provides a legal framework with the following definitions of biocidal actives and biocidal products based on those actives (author's emphasis applied):

"Article 3 (1): For the purposes of this Regulation, the following definitions shall apply:

(a) **'biocidal product'** means

- any substance or mixture, in the form in which it is supplied to the user, consisting of, containing or generating one or more active substances, with the intention of destroying, deterring, rendering harmless, preventing the action of, or otherwise exerting a controlling effect on, any harmful organism by any means other than mere physical or mechanical action,
- any substance or mixture, generated from substances or mixtures which do not themselves fall under the first indent, to be used with the intention of destroying, deterring, rendering harmless, preventing the action of, or otherwise exerting a controlling effect on, any harmful organism by any means other than mere physical or mechanical action.

(c) 'active substance' means a substance or a microorganism that has an action on or against harmful organisms".

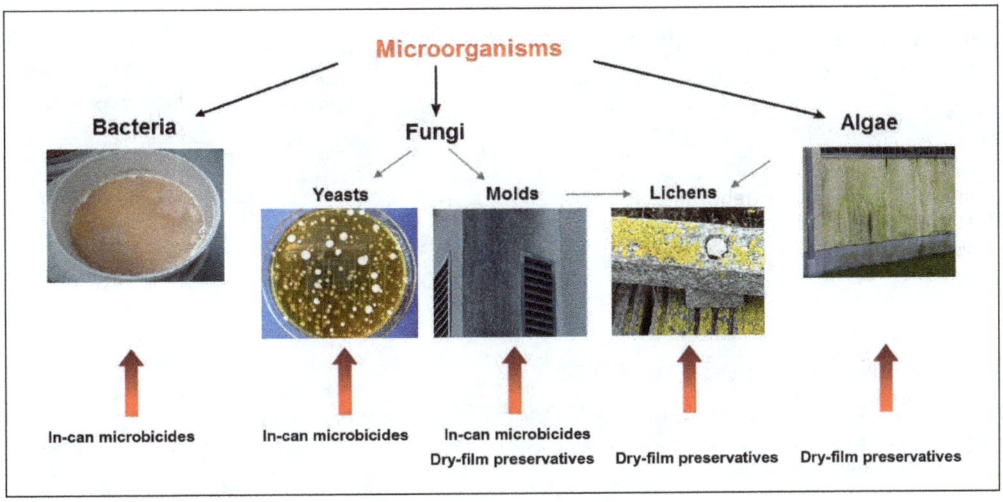

Figure 1.16: Sample schematic for the field of coating protection

Table 1.2: Grouping of frequently used microbicides by chemistry

Chemistry	Examples	Structure element
Alcohols	ethanol 1-propanol, 2-propanol benzyl alcohol	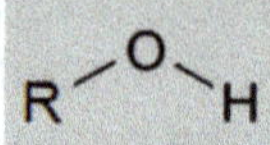
Aldehydes Di-aldehydes	formaldehyde (FA) glutardialdehyde (GDA)	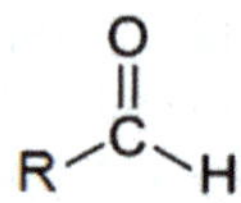
Formaldehyde releaser (O-Formal)	(ethylenedioxy)dimethanol (EDDM) dimethylformal (DMFL) benzylhemiformal (BHF)	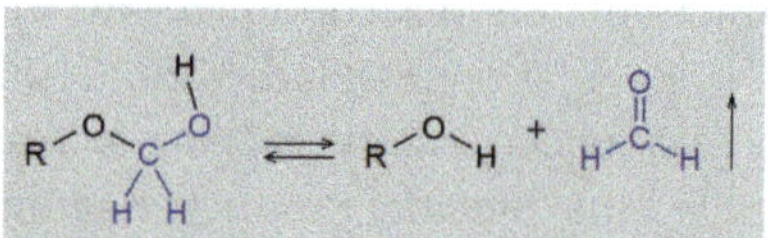
Formaldehyde releaser (N-Formal)	tetramethylol acetylenediurea (TMAD) 1,3-bis(hydroxymethyl)-5,5-dimethylhydantoin (DMDMH) hexahydro-1,3,5-tris (hydroxyethyl)-s-triazine (HHT)	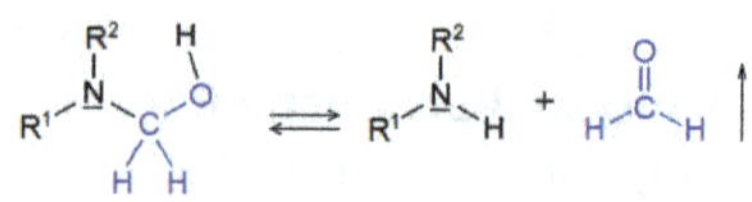
Phenolics Halogen phenoliccs	ortho-phenylphenol (OPP) 4-chloro-3-cresol (PCMC)	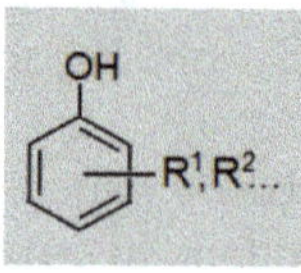
Carbamate derivatives Dithiocarbamate derivatives	3-iodo-2-propynyl butyl carbamate (IPBC) zinc bis(dimethyldithiocarbamate) (Ziram) bis(dimethylcarbamoyl) disulphide (Thiram)	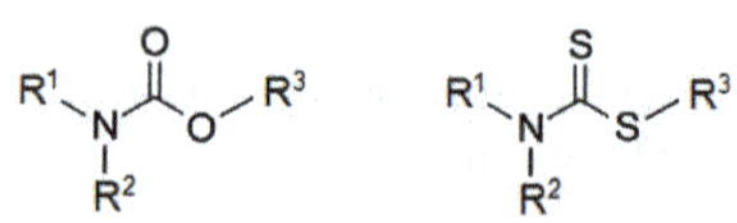
Pyridine-N-oxides	zinc pyrithione sodium pyrithione	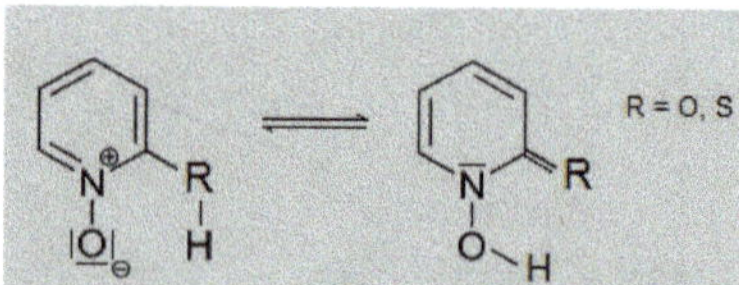
Azoles (1H-1,2,4 triazole derivatives)	tebuconazole propiconazole cyproconazole triadimefon	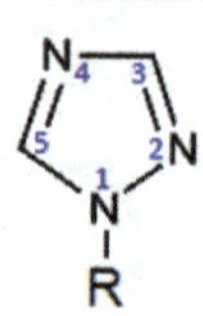
Benzimidazole derivatives Benzimidazole carbamate derivatives	thiabendazole (TBZ) carbendazim (BCM)	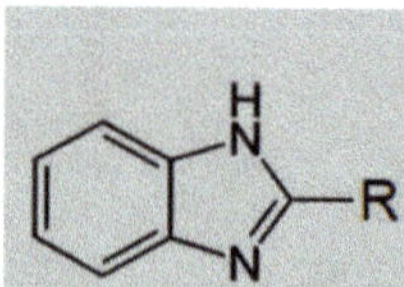

Table 1.2: Continue

Chemistry	**Examples**	**Structure element**
Isothiazolinone derivatives	5-chloro-2-methyl-4-isothiazolin-3-one (CMIT) 2-methyl-4-isothiazolin-3-one (MIT) benzisothiazolin-3-one(BIT) 2-octyl-4-isothiazolin-3-one (OIT) 4,5-dichloro-2-N-octyl-4-isothiazolin-3-one (DCOIT)	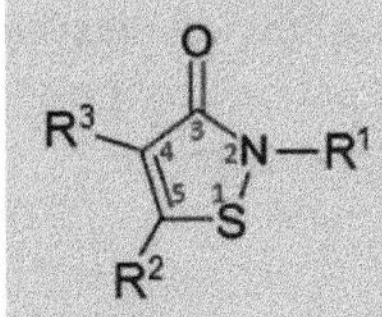
Compounds with activated halogens	2,2-dibromo-3-nitrilopropionamide (DBNPA) 2-bromo-2-nitropropane-1,3-diol (Bronopol, BNPD) 1,2-dibromo-2,4-dicyanobutane (DBDCB)	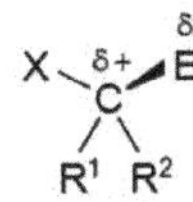
Quaternary ammonium compounds	alkyl dimethylbenzyl ammonium chloride (ADBAC) di-n-decyl-dimethylammonium chloride (DDAC) dimethyloctadecyl [3-(trimethoxysilyl)propyl] ammonium chloride	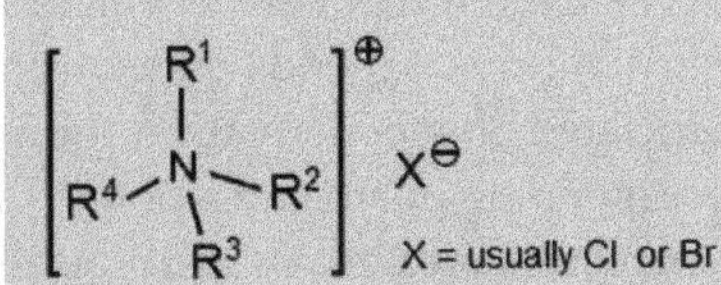
Triazine derivatives	terbutryne cybutryne	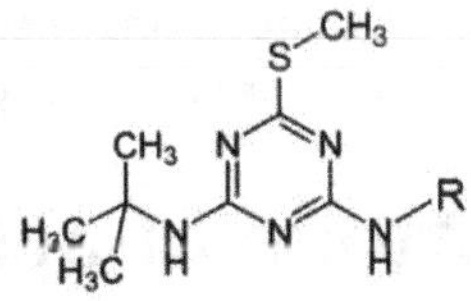
Phenylurea derivatives	3-(3,4-dichlorophenyl)-1,1-dimethylurea (Diuron) 3-(4-isopropylphenyl)-1,1-dimethylurea (Isoproturon)	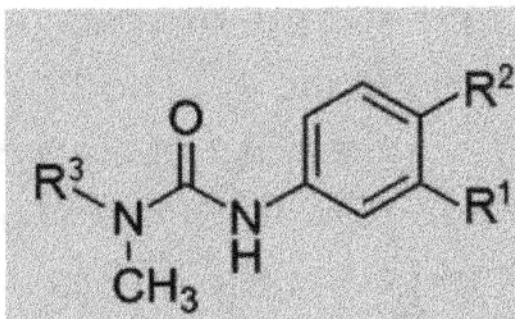
Pyrethroid derivatives	permethrin cypermethrin deltamethrin cyfluthrin	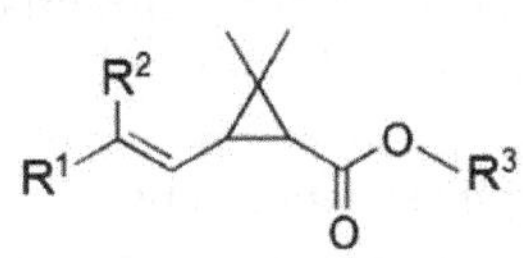
Silver compounds	silver zeolite compounds silver adsorbed on silicon dioxide	Ag, Ag^+
Diverse organic molecules	azoxystrobin (strobilurine derivative)	-

From a microbiological/chemical perspective, however, there are different ways in which to group microbicides. For instance, specialised literature[127, 128] suggests that microbicides used in the field of material protection be coarsely classified by the way in which microbial substances interfere with the target organisms. The majority of these chemical actives belong to at least one of the following classes[127]:
- electrophilically active compounds
- membrane-active substances
- chelate formers

Alternatively, microbicides can be grouped by their impact on the target organisms (e.g. bactericides, fungicides, algicides) and also by the intended application area (e.g. in-can microbicides and dry-film preservatives). A sample schematic for the field of coating protection is provided in Figure 1.16.

A further way to organise microbicides is to group them by chemical structure. Those structural elements which are known to exert an antimicrobial effect range from quite simple chemical molecules (e.g. lower primary alcohols) to complex chemical compounds, such as pyrethroid derivatives. Table 1.2 shows an overview of the most frequently employed antimicrobial chemistries in the field of material protection.

Finally, the modality of microbicidal interaction with microbial cells can vary substantially and often depends on environmental conditions, such as pH, and other factors. An undissociated phenol derivative, for instance, is much more effective against germs than is the corresponding phenolate anion[127]. Another example is the microorganisms' immediate environment: isolated germs are typically more sensitive to microbicides than those incorporated into a complex biofilm. Furthermore, it is often difficult to distinguish between a "characteristic" mode of action on a dedicated target site of a microbial cell and further consequences of that action. The underlying mechanisms can be quite complex and the biochemical processes in this context are not always fully understood (see also Section 1.3).

1.3 Mode of action of antimicrobial actives

Although a huge swathe of specialist literature is available that provides comprehensive information on the mode of action of antimicrobials and on the target sites of germs – see, for instance[127–140] – there is not always a consensus on the specifics of how microbicides interfere with the complex biochemical processes. However, it seems to be generally accepted that microbicides target multiple sites on or within the microbial cell and that it is the combined effects which ultimately lead to cell death[134].

Microbicides differ substantially in their antimicrobial activity. There are basically two different ways in which to affect microorganisms: either by growth inhibition (microbistatic effect) or by lethal action (microbicidal effect). While, for some applications (e.g. disinfection in hospitals), lethal action is the stated, preferred option, there are other applications in which preservatives are employed in near-lethal concentrations or even in sublethal doses (e.g. when a threshold for microbicide concentration triggers undesirable classification and labelling for the product requiring protection). The resulting microbistatic effect is typically reversible, unlike microbicidal action, which is final. An everyday example of the reversibility of a microbistatic effect is the freezing of food to protect it against spoiling. Although the freezing temperatures may kill some germs, others survive. However, the cold prevents the survivors from growing and multiplying as long as the temperature is kept low (microbistatic effect). Only when the food is again exposed to room temperature do the germs wake up and multiply.

Another example of the reversibility of microbistatic action in the field of material protection is the use of sublethal dosing with a microbicide to impair enzymatic activity in a microbial cell. As long as the active substance is present, microbistasis can be maintained. However, when the active begins to disappear (e.g. due to evaporation, degradation or consumption), the microbial cell has a chance to recover and to resume its disastrous action.

Basically, sublethal dosing harbours the risk of the emergence of microbial resistance, as the microorganisms in question become accustomed to the very chemicals designed to combat them. Consequently, any germs that have developed resistance will now withstand further use of such chemicals, with the result that in the worst case a certain chemical active might no longer serve its purpose. Development of resistance by microbial species can be highly problematic. This has become particularly apparent ever since the first reports were published in the field of health care about certain antibiotic-resistant bacteria called MRSA (methicillin-resistant *staphylococcus aureus*). MRSA is especially troublesome in hospitals, where patients with open wounds, invasive devices or weakened immune systems can come into contact with it. It can affect vital organs, lead to widespread infection (sepsis) and its resistance to several conventional antibiotics can even jeopardise the lives of patients who have just had a successful routine medical intervention.

The target site on a microorganism which the microbicide initially attacks is the microbial cell surface. A great many important biological processes take place there, and so there is plenty of scope for the active substance to interfere with the cell physiology. Although the outer cell wall, together with the cell membrane, essentially forms a barrier to the movement of chemicals across the channels (depending on their physical and chemical properties), transport of microbicides into the cell can be facilitated by special carrier molecules called chelating agents[127]. Once the cell interior has been invaded, the microbicides can then perturb the cell homeostasis (intracellular pH, membrane potential, solute transport) or influence vital cell functions, such as DNA, RNA and protein synthesis, enzyme-mediated reactions, intracellular energy transfer (ATP level), electron transport, and mitosis (division of the cell nucleus)[140]. Consequently, microbicidal attack can range from crude cell-membrane disruption to interference with more highly sophisticated microbiological cell functions. In a way, this can be likened to the different ways of stopping a wrist-watch. This can be done variously by removing the battery, by taking the balance wheel out of operation or by simply hitting the watch with a hammer (other approaches might be more appropriate for sundials).

Table 1.3 provides an overview of microbicidal chemistries and their major application areas. The latter were selected from the current so-called Article 95 list published by the European Chemicals Agency (ECHA) on its website[180]. This list is structured by active substance and includes all product types (PTs), i.e. the application areas for which a submission has been made under the BPR. Also highlighted are the respective microbicidal modes of action as well as the specific target sites in microbial cells, insofar as the very comprehensive literature sources allow of generalisation. The corresponding references are provided in the last column of Table 1.3.

Tabelle 1.3: Frequently used microbicides and their typical modes of action

Chemistry	**Important application areas***	**Mode of action/target sites/ important features**	**Ref.**
Alcohols	• skin/surface disinfection • cosmetics • health care • pharmaceuticals	–› membrane-active substances –› inhibition of DNA and RNA synthesis –› inhibition of cell wall synthesis (secondary effect) –› denaturation of proteins in the membranes of microbe cells –› dissolving membrane lipids –› not effective against spore –› fast acting in high concentrations, in particular when water is present	[127] [131] [141] [142]
Aldehydes and their releasers (FA-R)	• private area and public health area disinfectants • in-can protection of industrial fluids • preservatives for liquid-cooling and processing systems • slimicides • metalworking fluids	–› electrophilic active substances, alkylating agents –› can react with nucleophilic cell entities such as amino, amide and thiol groups –› inactivation of enzymes via reaction with the amino acids of their constituting proteins –› cross-linking of proteins, RNA, and DNA (formaldehyde) –› cross-linking of proteins in cell envelope and elsewhere in the cell (glutaraldehyde) –› effective against bacteria, fungi and spores (in particular glutaraldehyde) –› formaldehyde release ex N-formals slower than ex O-formals, i. e. less smell –› provide head space protection when used for in-can preservation	[127] [131] [143]
Phenolics	• disinfection in households and public buildings • animal stables • protection of plasterboards • paper industry • metalworking fluids • cooling waters and lubricoolants • fibre, leather, rubber and polymerized materials preservation • preservation of protein containing products (esp. PCMC)	–› membrane-active substances, membrane disruption –› adsorptive coverage of the microbe cell surface –› at higher concentrations cell wall attack and cell penetration –› reactions with the protoplasm and the cellular protein –› inhibition of enzyme activity –› general coagulation of cytoplasmic constituents –› undissociated phenol derivative more effective than phenolate anion –› broad spectrum efficacy against bacteria, fungi and yeasts –› at ambient temperature poor efficacy against bacterial spores –› antimicrobial effect of phenolics compromized in combination with detergent solutions due to formation of micelles (masking of the active)	[127] [130] [134] [136] [139] [144] [145]

Tabelle 1.3: Continue

Chemistry	Important application areas*	Mode of action/target sites/ important features	Ref.
Dithio-carbamate derivatives	• leather industry • household products • agriculture	–› conjugation with other molecules containing SH groups –› chelation with metal cations such as copper –› influence biological activities of different proteins, enzymes, and exert toxic effects –› dithiocarbamate anions highly reactive –› widely used as fungicides	[133] [146]
Carbamate derivatives	• wood protection (e. g. wood stains, cut/sawn timber) • dry film preservation • masonry protection • preservation of textiles and plastics • in-can protection of industrial fluids • metalworking fluids	–› microbicides of this class can widely differ in both efficacy and mode of action –› in fungi carbamates disturb cell membrane permeability and affect fatty acids metabolism (e. g. IPBC highly effective against blue stain fungi, sapstainers and wood rotting fungi)	[127] [129]
Pyridine-N-oxides	• dry film preservation • masonry protection • in-can protection of industrial fluids • metalworking fluids • cosmetics	–› membrane-active substances, chelating properties. –› influence on ATP levels, protein synthesis and nutrient transport –› spectrum of efficacy covers bacteria, yeasts and molds including alternaria –› not stable in stronger alkaline media (pH > 9.5) –› discoloring problems possible due to complexation with metals	[127] [147] [148] [149] [150]
Azoles (1H-1,2,4 triazole derivatives)	• wood protection (esp. against wood-decaying fungi) • dry film preservation • masonry protection • fibre, leather, rubber and polymerized materials preservation • agriculture	–› inhibition of ergosterol biosynthesis in fungal membranes –› at lower azole concentrations retardation of fungal growth –› at higher azole concentrations disruption of fungal membranes	[132] [135] [151] [152] [153]
Benzimidazole derivatives	• dry film preservation • masonry protection • preservation of textiles and plastics • wood protection • anthelmintic therapy (med.)	–› systemic fungicides –› selective action against fungal structural proteins (tubulin) –› inhibition of microtubules assembly –› interference with mitosis in fungi (division of the cell nucleus)	[154] [155] [156] [157]

Tabelle 1.3: Continue

Chemistry	Important application areas*	Mode of action/target sites/ important features	Ref.
Isothiazolinone derivatives	• in-can protection of industrial fluids • dry film preservations (OIT, DCOIT) • wood protection (OIT, DCOIT) • preservatives for liquid-cooling and processing systems • slimicide • metalworking fluids • antifouling products (DCOIT) • private area and public health area disinfectants	–› electrophilic active agents –› activated N-S bond reacts with nucleophils (e. g. amines or mercaptans) –› two-step mode of action: a) reaction with intracellular thiols of biocatalysts results in enzyme inhibition (minutes) b) destruction of protein thiols, generation of free radicals results in cell death (hours) –› halogenated derivatives additionally feature vinyl activated chloro atoms –› moreover, DCOIT contains an activated chloro atom in α-position to a carbonyl group –› broad spectrum of effectiveness, in particular against bacteria, fungi and yeasts –› water solubility depends on the structure (MIT > CMIT > BIT > OIT > DCOIT > BBIT)	[127] [158] [159]
Compounds with activated halogens	• in-can protection of industrial fluids • private area and public health area disinfectants • preservatives for liquid-cooling and processing systems • slimicides	–› electrophilic active substances –› electronegative groups in the microbicidal molecule create activated halogens –› reaction with nucleophilic entities such as thiols (e. g. in protein or enzyme structures) –› damage to membrane structures –› in particular effective against Gram-positive and Gram-negative bacteria	[134] [136] [160] [161]
Quaternary ammonium compounds	• human hygiene biocidal products • private area and public health area disinfectants • veterinary hygiene biocidal products • food and feed area disinfectants • protective agents for masonry • preservatives for liquid-cooling and processing systems • slimicide • embalming and taxidermist fluids	–› surface active agent –› membrane destabiliser –› damage the cytoplasmic membrane that controls the cell permeability –› at high concentrations cytoplasmic protein aggregation (loss of tertiary structure) –› in particular effective against bacteria, fungi, algae and lichens –› benzalkonium chloride solutions ideal for eliminating germs from coatings, stone etc. –› applicability limited due to possible foam problems & incompatibility with anionic compounds	[127] [131] [140] [143] [162] [163]
Triazine derivatives	• dry film preservation (algicide) • fibre, leather, rubber and polymerized materials preservation • protective agents for masonry (algicide)	–› interaction with the algal photosynthesis –› specific-binding site on a protein subunit of photosystem II –› no bactericidal or fungicidal properties, therefore usually combined with fungicides –› triazine derivative cybutryne not appoved for antifouling products due to Commission Implementing Decision (EU) 2016/107 from 27 Jan 2016	[133] [164] [165] [166] [167] [168]

Tabelle 1.3: Continue

Chemistry	Important application areas*	Mode of action/target sites/ important features	Ref.
Phenylurea derivatives	• dry film preservation (algicide) • protective agents for masonry (algicide)	–› interaction with the algal photosynthesis –› specific-binding site on a protein subunit of photosystem II –› no bactericidal or fungicidal properties, therefore usually combined with fungicides	[133] [165] [166] [167] [169]
Pyrethroid derivatives	• used as insecticides (esp. in the field of wood protection)	–› modify electrical activity in various parts of the nervous system –› stereoselective & structure-related interaction with sodium channels (primary target site) –› repetitive nerve activity, membrane depolarization & enhanced neurotransmitter release –› in wood preservation used for control of both beetles and termites –› low vapor pressure, high UV-resistance and low water solubility	[170] [171] [172] [173]
Silver compounds	• private area and public health area disinfectants • food and feed area disinfectants • drinking water disinfectants • film preservatives • fibre, leather, rubber and polymerized materials preservation • preservatives for liquid-cooling and processing systems	–› disruption of microbial cell wall and cell membrane –› influence on electron transport –› interaction with metabolic enzymes and nucleic acids –› low silver concentrations might result in formation of germ resistance –› high silver concentrations might cause discoloring problems	[174] [175] [176] [177] [178] [179]

* In the field of material protection the identified application areas were inter alia selected from the current so-called Article 95 list published by the European Chemicals Agency (ECHA) on its website. This list is structured per active substance and includes all product types (PTs) – i. e. the relevant application areas – for which a submission has been made for a particular substance. [180]

2 Coatings preservation

Microbicides are used in a variety of applications, including paints and coatings, stuccoes/plasters, polymer emulsions, adhesives and sealants, inks, plastics, wood treatments and many more. The coatings industry has to cope with a triangle of conflicting priorities between the various demands of technical solutions for contamination problems, increasingly stricter regulatory requirements, and commercial constraints. Regulations governing the production and application of microbicides affect the various markets in a number of ways, as they often restrict the use of certain microbicidal products while increasing the cost of time-consuming regulatory support. As a result, a number of microbicides are being withdrawn from some applications and traditional chemistries are being replaced by new products. While the regulations in Europe, America, Asia and other parts of the world are becoming more stringent, officials are agreed that these changes will benefit all stake-holders in the long term.

One of the biggest trends in the microbicide market is a desire for more eco-friendly products which have for example little or no VOC content and whose components exhibit minor ecological toxicity, and a desire for more cost-effective alternatives. This feeds directly into the need for a deep understanding of the intrinsic properties of microbicidal actives and the corresponding chemistries. Consequently, making the right selection from a range of legally accepted microbicides is key to protecting paints and coatings from microbial attack whilst simultaneously ensuring environmental sustainability. In this connection, the variety of potential chemical compounds is quite restricted and the proper allocation of products is becoming a more and more sophisticated task that requires a balancing act in terms of product design.

The extent to which the application of a microbicidal product poses a risk to human health and the environment depends on the manner of application. In general, risks can be reduced either by using a microbicide which has a low potential to cause harm or by applications that lead to low exposure levels.

Ideally, the microbicide should

- Offer a broad spectrum of activity against the target species
- Be stable, even under challenging environmental conditions (e.g. pH, UV light)
- Be compatible with the substrate to be protected
- Not pose a risk to human health
- Be environmentally sound
- Be cost-effective

Most commercially available microbicides do not meet all of these requirements. The choice of microbicidal product for a specific purpose often represents a compromise between expectations and reality. It is not always necessary to offer a full range that covers all species of microbes. Consequently, it is highly recommended that preliminary laboratory pre-trials be conducted on the suitability of the chosen microbicidal product for the intended purpose (see also Section 4).

All water-borne coatings typically contain organic components (e.g. binder, additives) which serve as nutrients for germs and are therefore susceptible to microbial degradation. In

Frank Sauer: Microbicides in Coatings

the absence of appropriate preservation in the wet state, the coatings become unusable and give rise to customer complaints. In addition, cured coatings come under microbiological attack from fungi, algae and lichens, particularly if the degree of atmospheric humidity is increased. Dry-film preservatives serve to prevent spoiling in and on the applied, cured paint film. Bio-deterioration of coatings is caused especially by the microbial groups illustrated in Figure 2.1.

These germs have only simple requirements for growth and proliferation, but these requirements are usually met during manufacturing processes, in storage, during transportation, and after application of water-borne products.

2.1 In-can preservation

Spoilage in the wet state is mainly due to bacterial and fungal growth. The product might develop a bad odour, discolouration, broken emulsions, a loss of viscosity or a decrease in pH. Evolution of carbon dioxide as a metabolite of microbiological action often causes the product containers to bulge until eventually the product is no longer sellable. The resulting economic loss is exacerbated by the ensuing need for waste disposal, which will incur additional costs.

Plant hygiene also requires the use of appropriate preservatives for the wet state (see Section 2.3)

The structural degradation of industrial fluids is usually attributable to complex microbial communities consisting of bacteria, fungi and yeasts. Ideally, for optimum efficacy, the microbicide will match the microbial species and its likelihood of occurrence in a given product. The following key factors should also be considered in the search to identify the best solution to a contamination problem:

- Type of product to be preserved
- pH & temperature at the point of dosage
- pH & temperature in the final product/in storage
- Typical or expected storage time of the product in question
- Requirements of specific regulations and approvals (e.g. eco-label, food contact)
- Exclusion of specific chemistries (e.g. sensitising agents)
- Currently employed microbicide(s), dosages and costs
- Known problems with microbicide(s) already in use
- Degree of service expected

Figure 2.1: Primary microorganisms responsible for the bio-deterioration of coatings

Table 2.1: EU work programme for the processing of applications under the biocides regulation[182]

Priority	Existing active substances for PT's	eCA to submit AR's and conclusions to ECHA	ECHA to start preparation of its opinion (submission to Commission max. 270 days later)
1st priority list	8, 14, 16, 18, 19, 21	31st Dez, 2015	31st Mar, 2016
2nd priority list	3, 4, 5	31st Dez, 2016	31st Mar, 2017
3rd priority list	1, 2	31st Dez, 2018	31st Mar 2019
4th priority list	6, 13	31st Dez, 2019	31st Mar, 2020
5th priority list	7, 9, 10	31st Dez, 2020	31st Mar, 2021
6th priority list	11, 12, 15, 17, 20, 22	31st Dez, 2022	30st Sep, 2023

AR = Assessment Report; eCA = evaluating Competent Authority; ECHA = European Chemicals Agency; PT = Product Type

Very often, different actives are combined in order to complement their microbicidal spectra (e.g. a very fast acting microbicide which kills the majority of germs in a short time frame, but which might be rapidly consumed due to its action, can be combined with a long-lasting microbicidal active which maintains protection during protracted storage of an industrial fluid). Moreover, standard laboratory test methods (see Section 4) are available that help to decide whether a selected antimicrobial product is appropriate for the intended purpose (Figure 2.2).

Although the initial Biocidal Product Legislation came into force at European level[181] back in 1998, even today, more than 18 years later, many actives for which an application has been filed are still awaiting final approval for the intended use – as indicated in the form of a so-called 'product type' (PT). Meanwhile, Commission Delegated Regulation (EU) No. 1062/2014 of 4th August 2014[182] foresees finalisation of the work programme in 2024, as shown in Table 2.1, i.e. some 26 years after the initial step.

However, some scepticism might be allowed as to whether even these significantly delayed timelines will be met.

The European Chemicals Agency (ECHA) regularly publishes on its website a list of biocidal active substances, also known as the 'Art. 95 List'[180], for which a dossier has been submitted by the relevant applicants, either under the former Biocidal Products Directive 98/8/EC ('BPD')[181], or under its successor, the Biocidal Products Regulation (EU) No. 528/2012 ('BPR')[126]. As

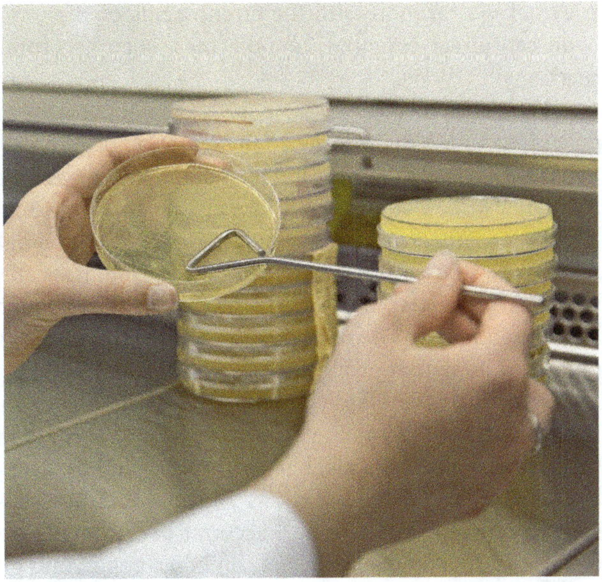

Figure 2.2: Standard laboratory test methods for choosing the right microbicidal products

of 9 May 2017, the Art. 95 List contained 52 different active substances for product type 6 (in-can preservation). Not all of them are utilised solely in the coatings area; some are also used in other areas, such as in-can preservation of glues and adhesives. The following sections discuss the most relevant and prominent actives for coatings designated for in-can preservation, grouped by chemistry.

2.1.1 Formaldehyde and formaldehyde-releasing compounds (FA-R)

Formaldehyde is an endogenous chemical compound that naturally occurs as a metabolite in humans and animals[183] and is also essential for the synthesis of amino acids, which are building blocks for important proteins in living creatures. Formaldehyde naturally appears in the troposphere as an intermediate oxidation product of hydrocarbons[184]. The latter are ultimately transformed into carbon monoxide, carbon dioxide, hydrogen and water. Finally, formaldehyde is naturally present in vegetables and is also formed in the early stages of the decomposition of plant residues in soil[184]. Given this, formaldehyde can be regarded as a chemical compound which is ubiquitous in the environment and which is additionally characterised by a very short half-life.

Regulation (EC) No. 1223/2009 of the European Parliament and of the Council on cosmetic products, the so-called 'Cosmetic Products Regulation'[185], allows the use of formaldehyde and selected formaldehyde releasers in cosmetics under specific conditions. According to this regulation, labelling of the finished cosmetic products is currently not required if the formaldehyde concentration does not exceed 500 ppm[185].

With regard to its mode of action, formaldehyde is an electrophilic substance and an alkylating agent (see also Table 1.3, Section 1). It can react with different nucleophilic elements of the cell, such as amino, amide and thiol groups, and is capable of inactivating enzymes via reaction with the amino acids of their constituent proteins. Finally, formaldehyde can affect essential biomolecules, such as proteins, RNA and DNA via cross-linking reactions[127, 131, 143].

According to the Art. 95 List[180], the active substance formaldehyde itself is not supported under the BPR for in-can preservation purposes (product type 6), but there are several applications in this regard for formaldehyde releasers under European biocides legislation. Formaldehyde releasers (FA-R) are chemicals which act as carrier systems for formaldehyde and are designed to release the inherently microbicidal active under use conditions after a time delay. In their undiluted form, they are quite stable but, in dilutions with water (e.g. in water-borne paints), they decompose more or less rapidly as a function of specific reaction kinetics and the molecular structure involved, as well as of further factors, such as pH, temperature and matrix properties[186].

Formaldehyde-releasing compounds vary substantially in physical appearance and chemical properties. They range from solid to liquid materials, from water-soluble to oil-soluble compounds, and have alkaline, neutral or slightly acidic properties. They enable formaldehyde to be used for applications which would otherwise be inaccessible to the active substance formaldehyde itself on account of its unfavourable properties[127]. As shown in Figure 2.3, formaldehyde releasers can be arranged into groups of O-formals and N-formals.

The bond strength of formaldehyde units is much greater in N-formals than in O-formals. Formaldehyde is therefore released from N-formals much more slowly than from

O-formals, and this leads to a considerably lower content of free formaldehyde in the corresponding formulations and subsequently in fewer undesirable side-effects, such as pungent odour or the need for classification and labelling. Figure 2.4 shows the situation for two examples of O-formals.

A set of active substances based on N-formal chemistry that is frequently encountered in coatings is illustrated in Figure 2.5.

Formaldehyde releasers are fast-acting bactericides/fungicides that are also efficacious against spores but whose effectivity spectrum against several fungi has some gaps. Thanks to their properties, they are very economical in use and are pH, temperature and redox stable. Formaldehyde itself is chemically reactive and so the active substance is consumed as it unfolds its microbicidal action. The typical dosage is within the range of 100 to 1000 ppm totally available formaldehyde content (see further). Figure 2.6 shows the stepwise release of formaldehyde, starting with the formaldehyde-releasing compound (ethylenedioxy)dimethanol (EDDM), in an industrial fluid.

1 mole (ethylenedioxy)dimethanol (EDDM) can theoretically liberate a maximum of 2 moles formaldehyde; the equilibrium of this reaction lies to the right. The percentage of formaldehyde in the molecule can be calculated as follows:

Molecular weight of (ethylenedioxy)dimethanol = 122.12 g/mol
Molecular weight of formaldehyde = 30.03 g/mol
Percentage of formaldehyde in the molecule = 2 * 30.03/122.12 * 100 = 49.2 %
i.e. nearly 50 % of the molecular weight of EDDM is potentially available formaldehyde.

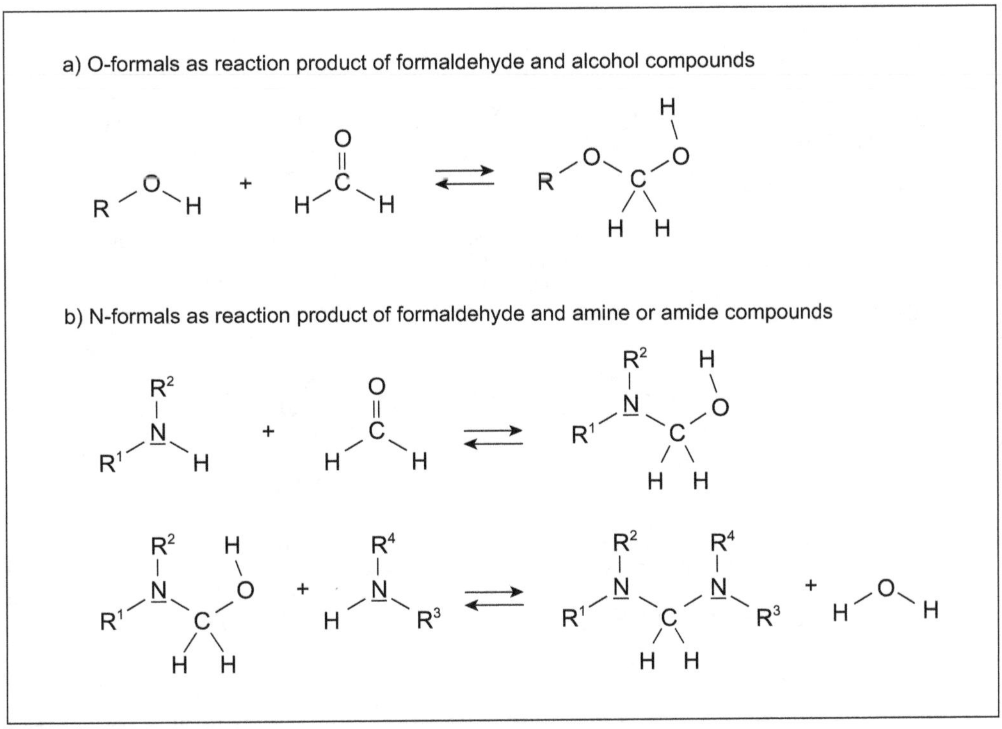

Figure 2.3: Schematic description of a) O-formal and b) N-formal constituents

The following calculation shows how the maximum available formaldehyde content in an industrial fluid to be protected is determined when a microbicidal product containing a formaldehyde releaser is added for in-can protection purposes. This value should not be confused with the free formaldehyde content in such a formulation, as there is always an equilibrium between the initial releaser compound and the corresponding reaction products. Consequently, the **free formaldehyde** content in a water-borne formulation is often substan-

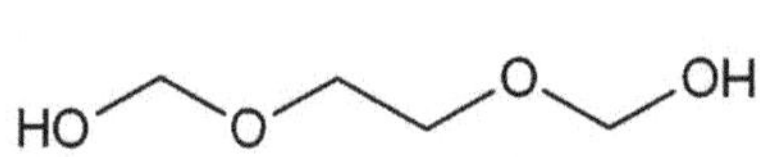

(ethylenedioxy)dimethanol
(EDDM)
CAS n° 3586-55-8

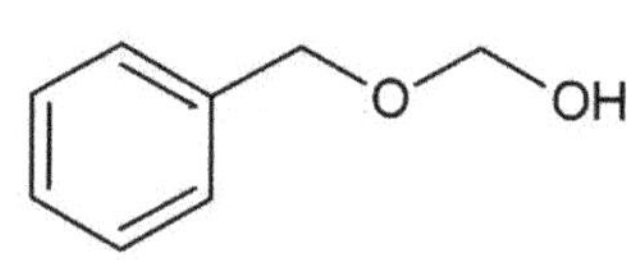

benzylhemiformal
(BHF)
CAS n° 14548-60-8

Figure 2.4: Examples of O-formals used in coatings (short forms in parentheses)

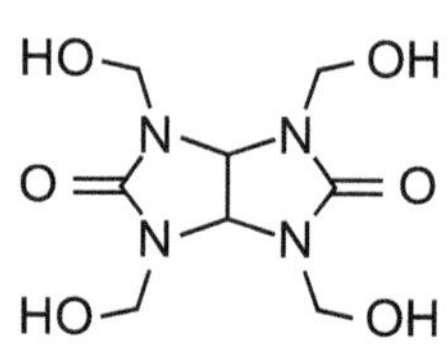

tetramethylol
acetylenediurea
(TMAD)
CAS n° 5395-50-6

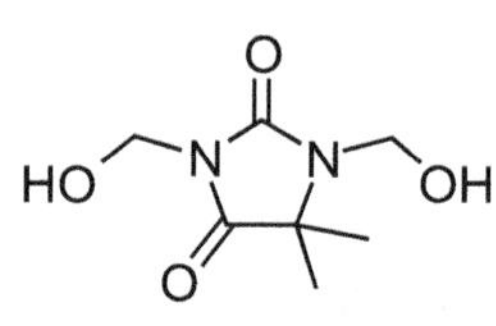

1,3-bis/hydroxymethyl)-
5,5-dimethylhydantoin
(DMDMH)
CAS n° 6440-58-0

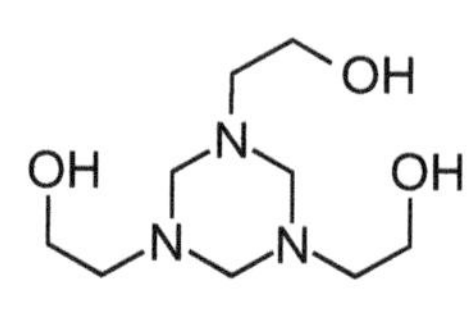

hexahydro-1,3,5-tris(hydroxyethyl)-
s-triazine
(HHT)
CAS n° 63310-09-8

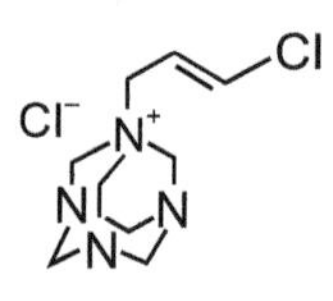

1-(3-chloroallyl)-3,5,7-triaza-1-
azoniaadamantane chloride
(CTAC)
CAS n° 4080-31-3

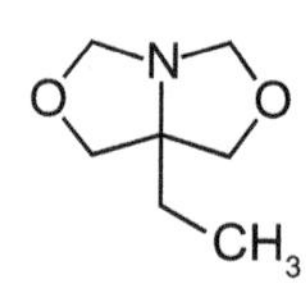

7a-ethyldihydro-1H,3H,5H-
oxazolo(3,4-c)oxazole
(EDHO)
CAS n° 7747-35-5

Figure 2.5: Examples of N-formals used in coatings (short forms in parentheses) tially lower than the maximum available formaldehyde content – the precise figure depends on the type of releaser used.

Assumptions:
30.0 % = 0.300 –› assumed EDDM content in the biocidal product formulation
0.2 % = 0.002 –› assumed dosage of the biocidal product formulation containing EDDM

Calculations:
49.2 % = 0.492 –› calculated FA content in the formaldehyde releaser EDDM (see above)
100 –› factor for conversion into percent
=› 0.002 * 0.300 * 0.492 * 100 ≈ 0.03 % (≡ 300 ppm) max. available formaldehyde content

In water-borne formulations preserved with O-formals, such as EDDM, the equilibrium lies typically more on the released side. By analogy with the calculation example above, this would mean that more than half of the calculated maximum available formaldehyde content does exist as free active (> 150 ppm). In contrast, the situation is reversed for N-formals where, with comparable amounts of totally available formaldehyde, noticeably lower values of free active are usually measured in the corresponding formulations (as per the above-mentioned example: roughly in the double-digit ppm range).

Since formaldehyde is consumed as it unfolds its microbicidal action, further active will be released in accordance with the respective equilibrium until the reservoir of releaser molecules is exhausted.

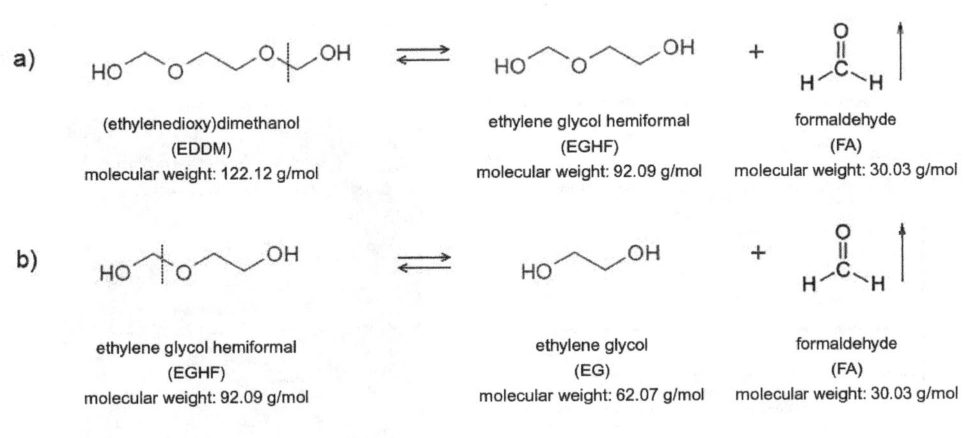

Figure 2.6: Stepwise release of formaldehyde, starting with the formaldehyde-releasing compound EDDM

A salient characteristic of this chemical family is its microbicidal activity in the headspace of a product container, owing to the volatility of formaldehyde (see Figure 2.7).

For instance, during storage of a paint can, water can evaporate from the liquid phase and condense again in the upper regions of the paint container, e.g. near the lid. In view of the fact that a paint can is not hermetically sealed (especially once it has been opened), it is basically possible for airborne germs to pass through the can closure and to settle down and proliferate in the condensed water drops (see Figure 2.7, left side). Frequent opening of the paint can during its shelf life affords further scope for germs to infiltrate the headspace. While microbicidal compounds are typically present in the main liquid phase, they may have a depleted presence in the headspace region. Consequently, the microbicidal activity in those areas might be insufficient to keep the paint container well preserved, even though the paint itself contains an effective microbicide.

While formaldehyde itself has been under discussion for quite some time on account of its possible adverse effect on human health, additional constraints have now been imposed on the various releasers as well. Since 1 January 2016, all products pursuant to Regulation (EC) No. 1272/2008 on the classification, labelling and packaging of substances and mixtures ('CLP Regulation') containing more than 0.1 % (1000 ppm) **free formaldehyde** have to be classified and labelled with "Carc. 1B, H350: May cause cancer"[187, 188].

2.1.2 Isothiazolinone derivatives

This class of electrophilic active agents belongs to the group of heterocyclic N, S compounds that may be regarded as derivatives of isothiazole (IUPAC name: 1,2-thiazole; CAS No. 288-16-4), a pseudo-aromatic ring system containing a sulphur and a nitrogen atom in direct adjacency (see Figure 2.8).

Their activated N-S bonds can react with different nucleophilic elements of the cell, such as amino, amide and thiol groups (see also Table 1.3, Section 1). In the first step, reaction

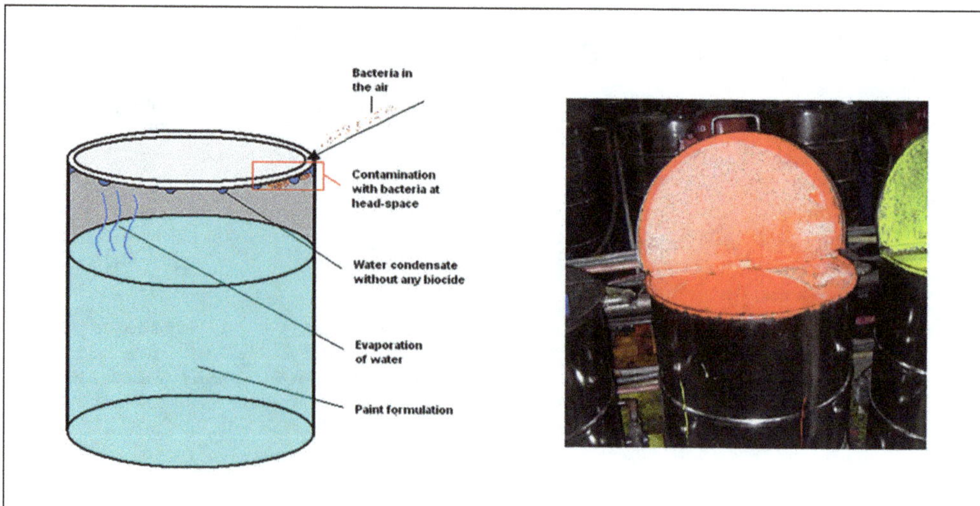

Figure 2.7: Headspace contamination in a paint container, left: schematic principle, right: example taken from practice

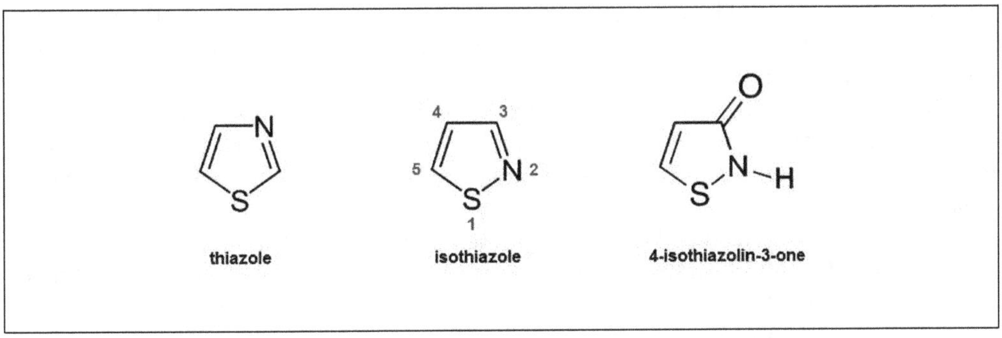

Figure 2.8: Structural relationship between isothiazolinone derivatives and the isothiazole backbone

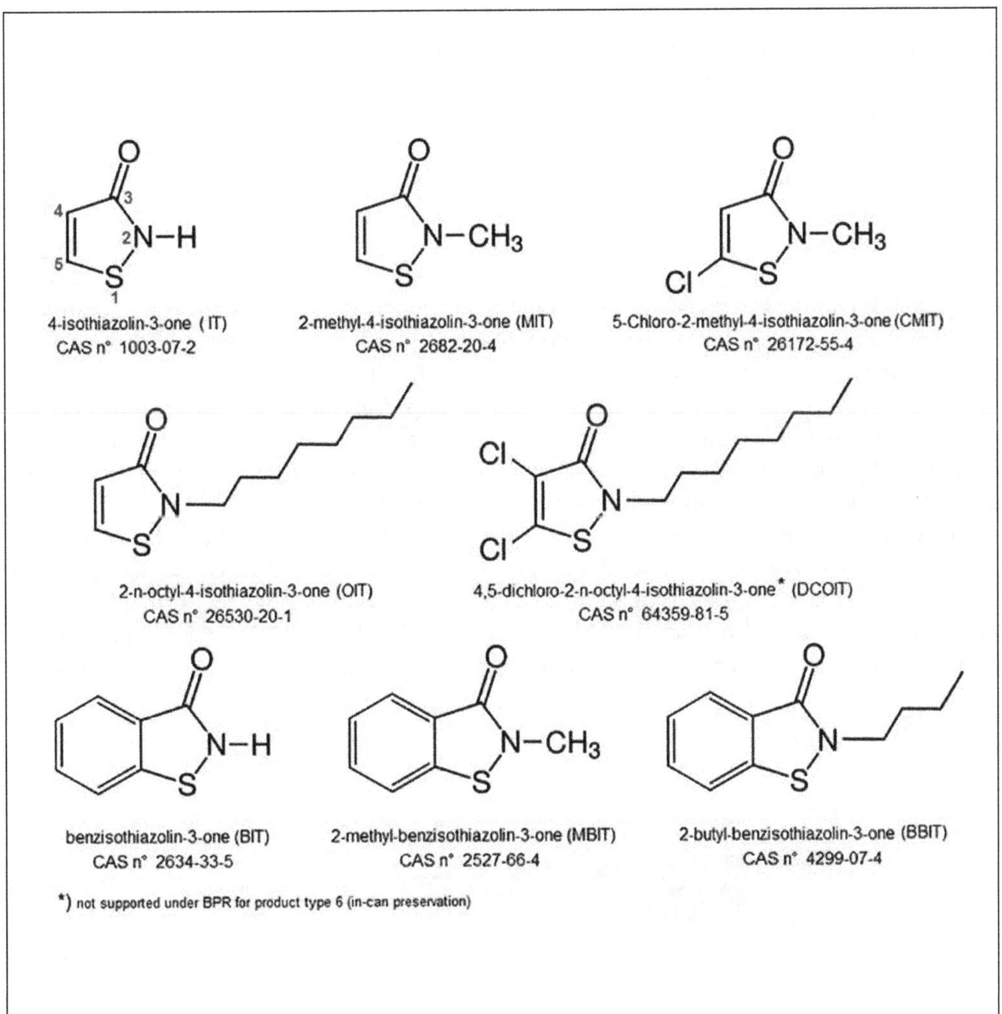

Figure 2.9: Overview of important isothiazolinone derivatives used in material protection (short forms in parentheses)

with intracellular thiols of biocatalysts leads to enzyme inhibition within a matter of minutes. Subsequently, destruction of protein thiols and generation of free-radicals eventually leads to cell death within hours[127, 158, 159].

Chlorinated isothiazolinone compounds additionally feature vinyl-activated halogens besides the reactive N-S bond, i.e. these contain two toxophoric structural elements in one molecule[189] and so they have much greater antimicrobial efficacy than the respective non-halogenated versions[190].

Isothiazolinone derivatives have manifold uses in material protection. One important application area of selected isothiazolinone derivatives is the preservation of industrial fluids in the wet state e.g.:

- In-can preservation
- Preservation of liquid-cooling and processing systems
- Preservation of metalworking fluids

But there are further fields of application where some isothiazolinone derivatives play an important role as well, such as:

- Disinfection in the private and public health spheres
- Use as slimicide
- Dry film preservation
- Wood protection
- Antifouling

Figure 2.9 provides an overview of important isothiazolinone derivatives that are commonly used in coatings protection (4-isothiazolin-3-one (IT) is only listed for the purpose of highlighting the basic structural element of this chemical family).

A key feature of in-can microbicides intended for preserving water-borne liquids is their solubility in water, as adequate solubility in water enables the microbicidal active to reach the place where it is actually needed. Figure 2.10 shows literature-based rankings of the relative solubility of isothiazolinone derivatives in water[127, 191–193].

It is not surprising that especially MIT, CMIT and BIT have long been utilised for in-can preservation purposes, whereas MBIT is relatively new on the scene[191]. Occasionally, the active substance OIT serves as a highly fungicidal combination partner for formaldehyde releasers or for the above-mentioned isothiazolinones CMIT, MIT or BIT intended for preserving industrial fluids, such as concrete additives, adhesives, starch slurries, cooling lubricants

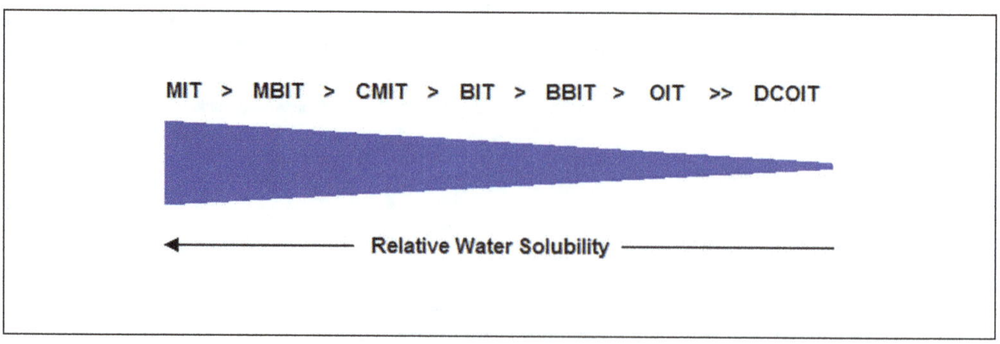

Figure 2.10: Relative solubility of isothiazolinone derivatives in water

and the like. The dichlorinated isothiazolinone derivative DCOIT is currently not supported under Biocidal Products Regulation[126] for product type 6, i.e. in-can preservation[180]. BBIT is advertised as being suitable for both in-can preservation and film preservation[194, 195].

One particular property of isothiazolinone derivatives is their sensitising effect on humans. Since 1 June 2015, where mixtures contain certain sensitising substances with a specific concentration limit (SCL) less than 0.1 %, the concentration limit for the hazard statement 'contains .... May produce an allergic reaction' (EUH208) is set at one tenth of this specific limit under the CLP Regulation[196]. Table 2.2 provides an overview of the different concentration limits of isothiazolinone derivatives associated with skin-sensitising properties.

As a result of the above-mentioned classification and labelling provisions, coatings products (such as paints) that are protected with higher amounts of isothiazolinone compounds may need to be labelled in accordance with one of the examples shown in Figure 2.11 (subject to additional labelling requirements triggered by possible further hazard criteria of the formulation components).

2.1.2.1 CMIT/MIT (3:1)

For in-can preservation purposes, the active 5-chloro-2-methyl-4-isothiazolin-3-one (CMIT; CAS No. 26172-55-4) is frequently combined with the active 2-methyl-4-isothiazolin-3-one (MIT; CAS No. 2682-20-4) in the ratio 3 : 1 (Figure 2.12).

This active combination in the stated ratio above is so well established on the market that it has its own CAS No. (55965-84-9). The mixing ratio 3 : 1 represents the optimum for microbicidal efficacy.

This particular combination is also registered under the Cosmetic Products Regulation[185] for rinse-off products at concentration levels below 0.0015 % (15 ppm).

CMIT/MIT formulations in coatings are, to an extent depending on the matrix properties, typically stable in a pH range of approx. 2.5 to 9 and at temperatures below approx. 60 °C. Water-soluble Mg^{2+} or Cu^{2+} salts in the biocidal product formulation can enhance the stability of the actives. Nucleophilic compounds in their non-protonated form, such as amines or

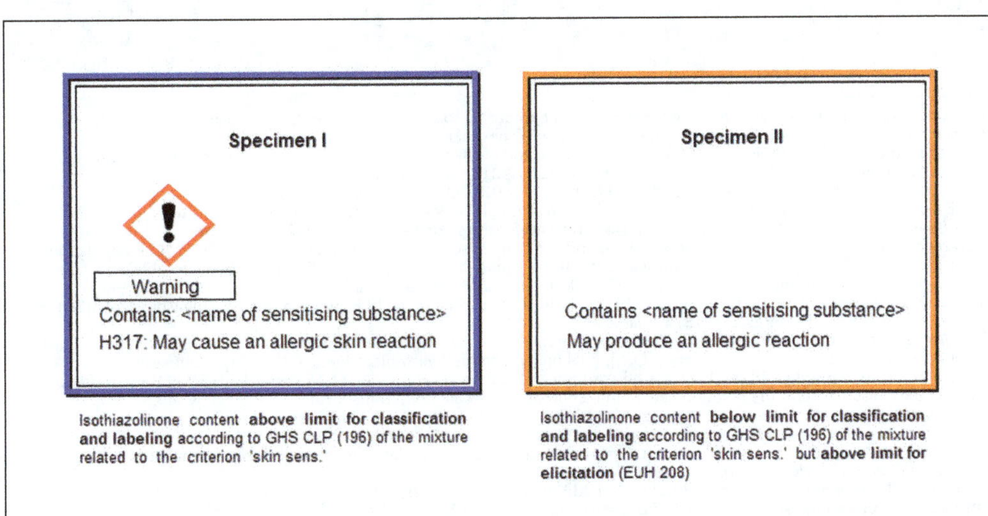

Figure 2.11: Sample labelling for formulations containing isothiazolinone (IT) derivatives

Table 2.2: Concentration limits for the classification and labelling of mixtures containing isothiazolinone derivatives associated with the criterion 'skin sensitising'

			CL[1]		EUH208		
Active(s)	**CAS n°**	**Criterion**	**[% w/w]**	**[ppm]**	**from [ppm]**	**to [ppm]**	**Remark**
CMIT/ MIT (3:1)	55965-84-9	skin sens. 1, H317	≥ 0.0015	≥ 15	1.5	15	current entry in Annex VI of CLP Regulation[2]
		skin sens. 1, H317	≥ 0.1000	≥ 1000	100	1000	CEPE Guidance (2014)[3]
MIT	2682-20-4	skin sens. 1A, H317	≥ 0.0600	≥ 600	60	600	proposed future entry in Annex VI of CLP Regulation[4]
		skin sens. 1A, H317	≥ 0.0015	≥ 15	1.5	15	DEPA Proposal Aug 2015[5], see also SCCS Opinion[6]
BIT	2634-33-5	skin sens. 1, H317	≥ 0.0500	≥ 500	50	500	current entry in Annex VI of CLP Regulation[2]
MBIT	2527-66-4	skin sens. 1A, H317	≥ 0.1000	≥ 1000	100	1000	proposed future entry in Annex VI of CLP Regulation[7]
BBIT	4299-07-4	skin sens. 1, H317	≥ 1.0000	≥ 10000	1000	10000	current entry in Annex VI of CLP Regulation[2]
OIT	26530-20-1	skin sens. 1, H317	≥ 0.0500	≥ 500	50	500	current entry in Annex VI of CLP Regulation[2]
		skin sens. 1A, H317	≥ 0.0050	≥ 50	5	50	proposed future entry in Annex VI of CLP Regulation[8]
DCOIT	64359-81-5	skin sens. 1, H317	≥ 0.0250	≥ 250	25	250	CEPE Guidance (2014)[3]
		skin sens. 1A, H317	≥ 0.0010	≥ 10	1	10	proposed future entry in Annex VI of CLP Regulation[9]

[1] CL (concentration limit): either generic concentration limit according to Annex I, Part 3 of CLP Regulation[2] or specific concentration limit according to Annex VI of CLP Regulation[2] or others

[2] Regulation (EC) No 1272/2008 of the European Parliament and of the Council on classification, labelling and packaging of substances and mixtures (consolidated version dated 01.01.2016); available at http://eur-lex.europa.eu/legal-content/EN/TXT/PDF/?uri=CELEX:02008R1272-20160101&rid=1

[3] CEPE Guidance on labelling decorative paint with skin sensitizing biocides - Final (28.10.2014); available at: http://sveff.se/files/2015/04/CEPE-guidance-on-labelling-with-biocides-skin-sensitizers-final-28-Oct-14.pdf

[4] CLH report, Proposal for Harmonised Classification and Labelling based on Regulation (EC) No 1272/2008 (CLP Regulation), Annex VI, Part 2; Dossier submitter: Chemicals Office of the Republic Slovenia, July 2015; information available at: http://echa.europa.eu/documents/10162/0d4c2335-6009-4e65-9278-c7f10a3a2dad

[5] The Danish Environmental Protection Agency (DEPA), Danish Comments to the CLH proposal for 2-methylisothiazol-3(2H)-one (MIT), (CAS no 2682-20-4), 28.08.2015; available at: http://echa.europa.eu/documents/10162/13626/clh_2-methylisothiazol_attachment_03_en.pdf

[6] Scientific Committee on Consumer Safety (SCCS), final opinion on methylisothiazolinone (MI) (P94) - Submission III (SCCS/1557/15), December 2015: "For rinse-off cosmetic products, a concentration of 15 ppm (0.0015%) MI is considered safe for the consumer from the point of view of induction of contact allergy"; available at: http://ec.europa.eu/health/scientific_committees/consumer_safety/docs/sccs_o_178.pdf

[7] Submitted CLH proposal; Dossier submitted by Poland (December 2015); available at: http://echa.europa.eu/de/registry-of-submitted-harmonised-classification-and-labelling-intentions/-/substance-rev/12105/del/50/col/synonymDynamicField_235/type/desc/pre/1/view

[8] Submitted CLH proposal; Dossier submitted by United Kingdom; (December 2015); available at: http://echa.europa.eu/registry-current-classification-and-labelling-intentions/-/substance-rev/9772/term

[9] Submitted CLH proposal; Dossier submitted by Norway; Expected date of submission: February 2016; available at: http://echa.europa.eu/registry-current-classification-and-labelling-intentions/-/substance-rev/12435/term

thiols, can deactivate the electrophilic isothiazolinones. Additionally, the latter are susceptible to oxidising and reducing agents[127]. It therefore makes sense to first check their efficacy and service life in benchmark tests (see also Section 4) – this approach is generally advisable for all microbicides used in the field of material protection.

Microbicidal products containing CMIT/MIT (3 : 1) are very economical in use and provide an excellent spectrum of antimicrobial activity against bacteria (gram-positive as well as gram-negative), fungi and yeasts. Very often, formulations containing the active substance combination CMIT/MIT (3 : 1) are offered on the market in such delivery forms that a typical recommended dosage of the respective microbicidal product (usually 0.1 % to 0.2 %) yields a final active substance content of no more than 0.0015 % (15 ppm) in the product to be protected, i.e. just below the classification and labelling limit for this combination as "Skin Sens. 1; H317: May cause an allergic skin reaction" (see also Table 2.2).

Sample calculation
Assumptions:
1.5 % = 0.015 –› assumed CMIT/MIT (3 : 1) content in the microbicidal product
0.1 % = 0.001 –› assumed dosage of the microbicidal product containing CMIT/MIT (3 : 1)

Calculations:
100 –› factor for conversion into percent
=› 0.015 * 0.001 * 100 = 0.0015 % (≡ 15 ppm)
calculated CMIT/MIT (3 : 1) content in the product to be protected.

2.1.2.2 MIT

The active 2-methyl-4-isothiazolin-3-one (MIT; CAS No. 2682-20-4) (Figure 2.12) is also authorised under the Cosmetic Products Regulation[185] in ready-for-use preparations at concentration levels below 0.01 % (100 ppm). But authorities and scientific committees have al-

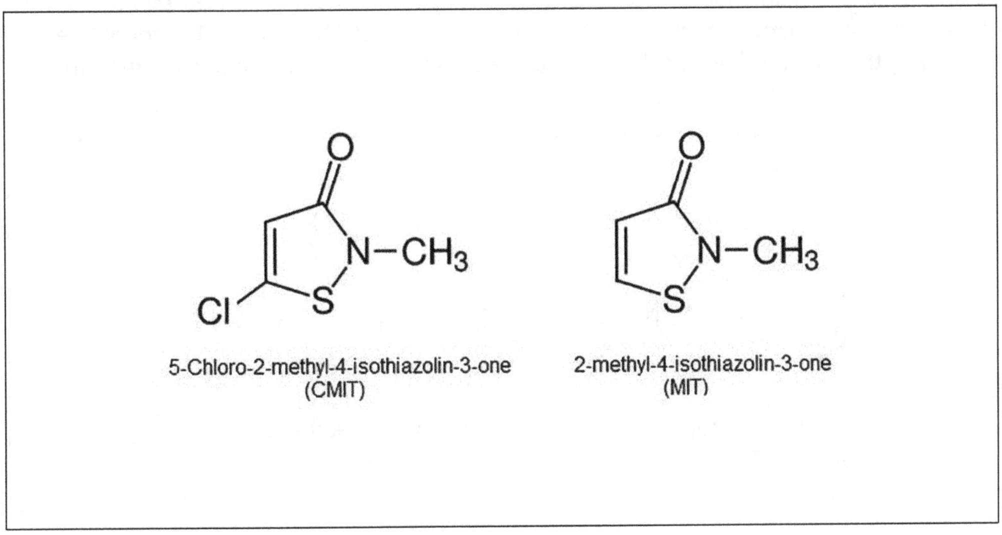

Figure 2.12: Molecular formulas of CMIT and MIT

ready submitted proposals to set a limit for classification as skin sensitiser (Skin Sens. 1A; H317) at 15 ppm due to the properties of this active (see Table 2.2).

Until now, exceeding the concentration level of 100 ppm in coatings would trigger the GHS hazard statement 'EUH208' (contains 2-methyl-4-isothiazolin-3-one. May produce an allergic reaction) while concentrations higher than 1000 ppm would entail classification and labelling as "Skin Sens. 1; H317: May cause an allergic skin reaction" including the corresponding warning symbol pursuant to the CEPE Guidance on labelling decorative paint with skin-sensitising biocides[197]; (see also Table 2.2).

MIT is a relatively slow-acting microbicide that has good efficacy against bacteria, but its activity spectrum has some gaps against certain fungi. It is therefore often combined with other actives, such as formaldehyde releasers or BIT (see below). The typical dosage for in-can preservation is 100 to 200 ppm. Like CMIT/MIT formulations, it is susceptible to oxidising and reducing agents as well as to attack by nucleophiles, such as amines and thiols. The aqueous solution of this active features good stability over a relatively wide pH range (2 to 10) and at elevated temperatures (up to approx. 80 °C), along with excellent solubility in water. Further advantages of the active are its availability as a salt-free formulation and the absence of halogens in the molecule, as a result of which it does not contribute to the AOX value (a measure of the adsorbable organic halogen content as a key factor in the assessment of the quality of water and sewage sludge).

2.1.2.3 BIT

Unlike CMIT and MIT, the active benzisothiazolin-3-one (BIT; CAS No. 2634-33-5) (Figure 2.13, left) is not authorised under the Cosmetic Products Regulation[185].

Exceeding the concentration level of 50 ppm in coatings would trigger the GHS hazard statement 'EUH208' (contains benzisothiazolin-3-one. May produce an allergic reaction) while concentrations higher than 500 ppm would entail classification and labelling as "Skin Sens. 1; H317: May cause an allergic skin reaction" including the corresponding warning symbol according to CLP Regulation[196] (see also Table 2.2).

BIT is a slow-acting bactericide with an unbalanced efficacy spectrum that has gaps, especially against *Pseudomonas spp.* and several fungi. It is therefore often combined with other microbicidal actives, such as formaldehyde releasers, (CMIT)/MIT or Bronopol (see Section 2.1.3) that can fill these gaps. The typical dosage of BIT for in-can preservation in coat-

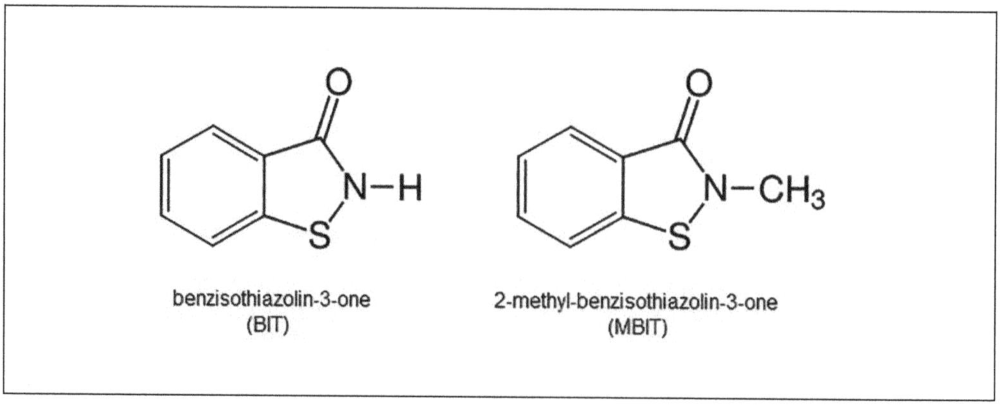

Figure 2.13: Molecular formulas of the benzisothiazolinone derivatives BIT and MBIT

ings is 100 to 500 ppm. The presence of alkalis and amines leads to the formation of water-soluble salts. BIT itself is highly soluble in organic solvents, such as alcohols and glycols[127]. The active BIT is also used in wet-state preservation as a combination partner for the active substance zinc pyrithione (ZnP, IUPAC name: zinc bis(2-thioxopyridin-1(2H)-olate), CAS No. 13463-41-7, see also Section 2.2.1). Zinc pyrithione is a dedicated fungicide but has bactericidal properties as well and could therefore cover the gaps left by BIT against certain bacteria and fungi. Examples of those formulations for in-can preservation are mixtures of BIT/ZnP in the ratio 1 : 1 and 1.5 : 1. The dosage of BIT in this case is 14 to 300 ppm, while that for zinc pyrithione is 14 to 200 ppm. The active zinc pyrithione will be discussed further in Section 2.2.1 (Fungicides for coatings protection).

A salient characteristic of BIT is its excellent thermal and chemical stability, as a result of which it can be employed over a wide pH range (up to pH 14) and also at high temperatures (at least 100 °C), a fact which supports broad, versatile use. It can often be added at an early stage in the production process to effect preservation from the outset. Given the high heat resistance and the low volatility of BIT, loss of this active during production, even upon subsequent heating, is generally not to be expected. Like the isothiazolinones described above, BIT is susceptible to oxidising and reducing agents, such as persalts, sulphites and dithionites[127]. Owing to the absence of halogens in the molecule, BIT does not contribute to the AOX value.

2.1.2.4 MBIT

The active substance 2-methyl-benzisothiazolin-3-one (MBIT; CAS No. 2527-66-4) is described in the literature[191] as being generated from dithio-2,2'-bis-benzmethylamide (DTBMA; CAS No. 2527-58-4) in alkaline media via hydrolysis (Figure 2.14). In this context, it has been reported that the conversion from DTBMA to MBIT can be fairly quantitative in solvency-moderating matrixes[198].

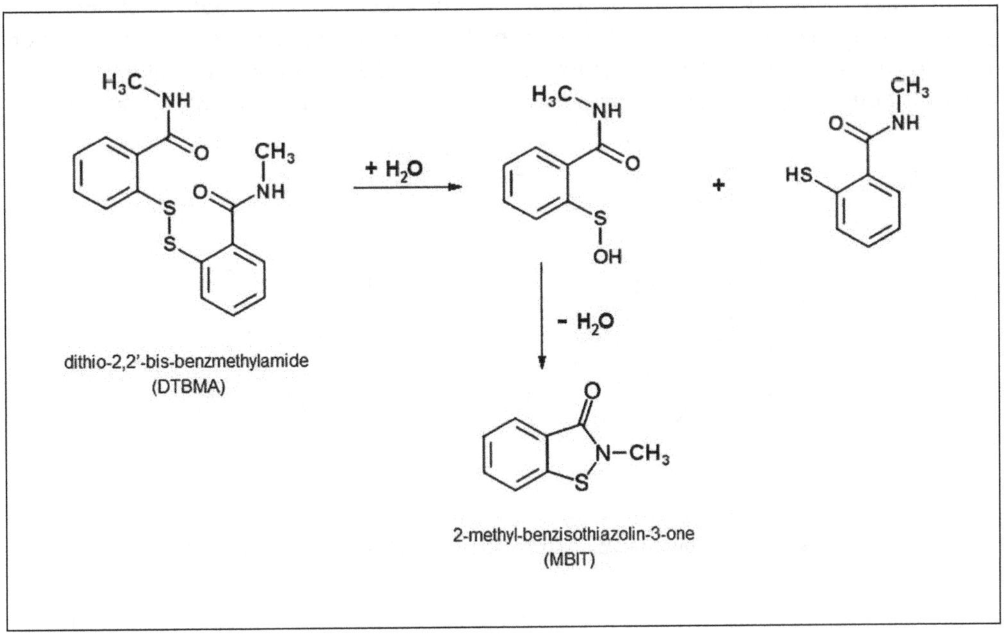

Figure 2.14: Conversion of DTBMA to MBIT via hydrolysis

DTBMA itself is an established microbicidal active that has long been used in material protection and is also included in the current Art. 95 List published by ECHA[180] for product type 6 (in-can preservation). MBIT also is included in that list as a 'new active' for product type 6 (in-can preservation) and product type 13 (metalworking fluids). However, a recently published draft Commission Implementing Decision[315] suggests non-approval of 2-methyl-1,2-benzisothiazol-3(2H)-one as an active substance for use in biocidal products of product-type 13 due to identified unacceptable risks for the environment (groundwater) for this use.

The solubility of MBIT in water (approx. 2 %) is greater than that of non-deprotonised BIT (approx. 0.1 %) and is therefore said to be more available in the aqueous phase to combat microorganisms[191]. In addition, synergistic effects have been reported with other microbicidal actives, especially with MIT, that might close some performance gaps of the latter. Its pH stability is cited as being 2 to 10[198]. Owing to the absence of halogens in the molecule, MBIT does not contribute to the AOX value.

Like other isothiazolinones, MBIT is reported to be a sensitising substance. According to the classification and labelling proposal submitted by the Polish Competent Authority[199] in December 2015, exceeding the concentration level of 100 ppm in coatings would trigger the GHS hazard statement 'EUH208' (contains 2-methyl-benzisothiazolin-3-one. May produce an allergic reaction) while concentrations higher than 1000 ppm would entail classification and labelling as "Skin Sens. 1; H317: May cause an allergic skin reaction" including the corresponding warning symbol (see also Table 2.2).

In the field of coatings, the isothiazolinone derivatives OIT, DCOIT and BBIT (see Figure 2.9) are typically discussed in the context of preservation of materials against (film) fungi. They will therefore be described in Section 2.2 (dry-film preservation).

2.1.3 Compounds with activated halogens

Microbicides in this class are electrophilic active substances containing both electronegative groups and halogens in the same microbicidal molecule. Halogens, such as those in α-position or in vinyl position to the electronegative group (Figure 2.15), can be activated by the electron distribution in the respective compound[127]. These substances enter into a reaction with

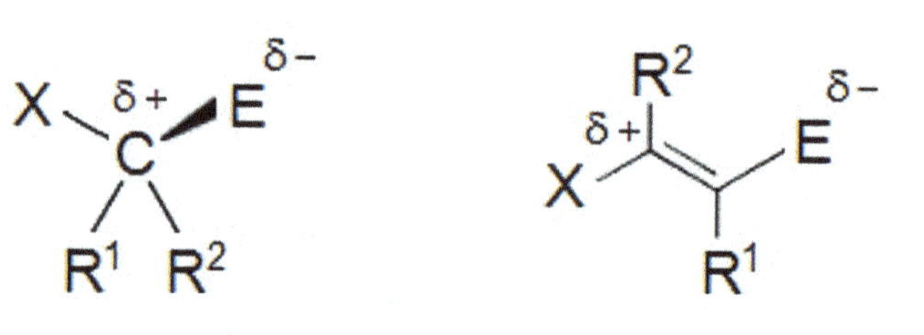

Figure 2.15: Schematic illustration of organic compounds with activated halogens

nucleophilic cell entities, such as thiols (e.g. in protein or enzyme structures), that ultimately leads to cell membrane damage. Those actives are especially effective against Gram-positive and Gram-negative bacteria[134, 136, 160, 161].

Figure 2.16 shows frequently used in-can preservatives from the class of compounds containing activated halogens.

2.1.3.1 Bronopol

The active substance 2-bromo-2-nitropropane-1,3-diol (BNPD; CAS No. 52-51-7) (Figure 2.16) reportedly creates disulphide cross-links at the cell wall by interfering with oxidation of the amino acid cysteine (contains the -SH moiety) to cystine derivatives (which contain R-S-S-R moieties) under aerobic conditions. The formation of microbicidal active oxygen species has also been discussed in this context. The combined action is reported to lead to cell death[134, 136, 160, 161, 200]. Under anaerobic conditions, the electrophilic microbicidal active converts nucleophilic cellular and extracellular moieties and so affects important microbial structures, such as proteins and enzymes[127].

Bronopol is authorised under the Cosmetic Products Regulation[185] in ready-for-use preparations at concentration levels below 0.1 % (1000 ppm). The active is susceptible to high pH and elevated temperatures (–› decomposition) as well as to ammonia and amines. It exhibits good solubility in water (approx. 25 %) and in various organic solvents. Its antimicrobial efficacy is not usually impaired by anionic nor non-ionic surfactants. The typical dosage for in-can preservation is 50 to 300 ppm (active substance). Like all bromine containing compounds, bronopol will contribute to the AOX value.

Bronopol is generally not considered to be a fast-acting microbicide. It is effective against gram-positive and gram-negative bacteria, especially *Pseudomonas spp.* against which, for example, the isothiazolinone BIT is inefficacious. Its activity against moulds and yeasts is unincisive and so it is advisable to combine bronopol with other actives in the event that a distinct fungicidal effect is needed.

It is worth mentioning that this active also covers sulphate-reducing bacteria (SRB). These microbial species are ubiquitous in anaerobic environments, where they are involved in the degradation process of organic compounds. Additionally, they take part in the sulphur and

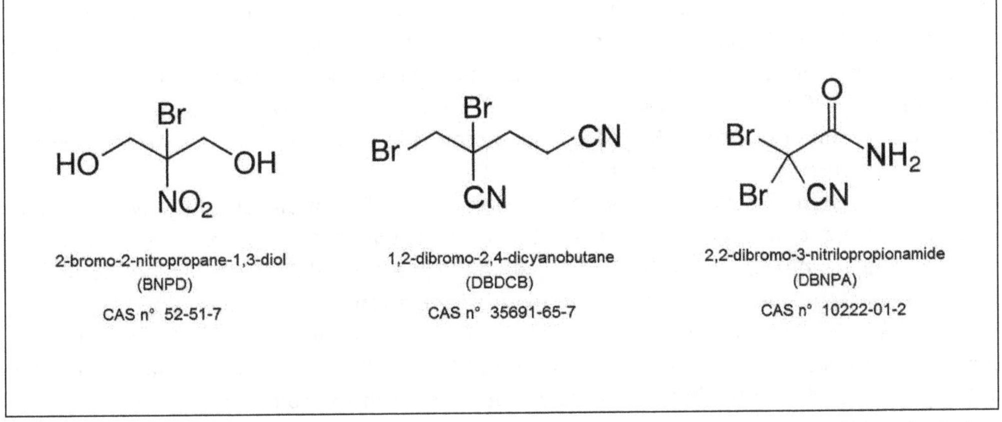

Figure 2.16: Frequently used in-can preservatives from the class of compounds containing activated halogens

carbon cycles[201]. Due to the generation of hydrogen sulphide as metabolite, sulphate-reducing bacteria can be responsible for corrosion of metals, a factor that merits thorough consideration should SRBs be detected in industrial facilities.

Common methods for the synthesis of bronopol are based on the condensation reaction of nitromethane with formaldehyde in alkaline aqueous medium, followed by bromination of the intermediate, which is usually not isolated under commercial technological conditions[201]. At first sight, it does not seem implausible that a reverse reaction might release formaldehyde molecules under certain conditions. For this reason, bronopol is occasionally cited in literature as a formaldehyde releaser (see, e.g.[202, 203]), especially in alkaline media. However, during a workshop on formaldehyde releasers held in Warsaw on 14th to 15th January 2008 with representatives from the Competent Authorities in Poland, Spain, Austria, Germany, Greece and Hungary, there was a discussion on whether or not the activity of bronopol is actually contingent on formaldehyde release. It was stated that the microbicidal action is actually based to a greater extent on the reaction of the bronopol molecule with thiol groups in microbial cells and the formation of oxygen radicals[204]. This opinion was agreed by the experts of all participating countries at that meeting and a proposal was also made to exclude this active substance from the group of formaldehyde releasers[204]. In September 2014, a further document was published[205] arising from a meeting of representatives of Competent Authorities for the implementation of Regulation 528/2012 (BPR) concerning the management of active substances generated *in situ*. In this document, bronopol is no longer listed under the category 'formaldehyde releaser' but is now found under 'miscellaneous substances supported under the review programme'. This view is also supported by other references which attribute the microbicidal effect of bronopol to mechanisms other than formaldehyde release[206, 207]. Furthermore, previously unreleased laboratory chamber tests and headspace analyses with standard decorative paint formulations containing 100 to 420 ppm bronopol revealed evidence that the test material gave rise to merely negligible amounts (< 1 ppm) of formaldehyde[208, 209]. It can therefore be hypothesised that, under standard processing conditions, substantial release of formaldehyde is not to be expected from coatings containing bronopol.

2.1.3.2 1,2-Dibromo-2,4-dicyanobutane (DBDCB)

The active substance 1,2-dibromo-2,4-dicyanobutane (DBDCB; CAS No. 35691-65-7) (Figure 2.16) reacts, like bronopol, with nucleophilic groups in the microbial cell. In waterborne formulations, the substance is susceptible to high pH (> 9) and elevated temperatures (> 60 °C) but remains stable under the usual processing conditions for coatings. The best performance of DBDCB is achieved at pH 4 to 8[127]. Its solubility in water at ambient temperatures is quite limited (approx. 0.38 %) and so microbicidal product formulations are typically offered in the form of aqueous dispersions.

Iron oxide pigments can possibly interfere with the DBDCB molecule, thereby potentially necessitating higher dosage due to inactivation of the active. In addition, a certain risk of discolouring (yellowing) which has been reported in the presence of iron ions can be avoided by adding sequestrants (e.g. EDTA) to the aqueous formulation to be protected[127].

DBDCB provides a broad spectrum of microbicidal activity against bacteria, fungi and yeasts. The usual dosage for in-can protection is 250 to 1000 ppm (active substance). Thanks to the presence of bromine in the molecule, DBDCB contributes to the AOX value.
The Implementing Regulation for approval of DBDCB in product type 6 (in-can preservation) was published in July 2016 on the ECHA website[210].

Table 2.3: Important properties of prominent in-can microbicides used in the coatings area

Active ingredient	Efficacy Bacteria	Fungi	Yeast	Reaction time	Stability[(1)] pH	T [°C]	Head space protec.	AOX	Typ. dosage [ppm]
CMIT/MIT (3:1)	++	++	++	rapid	2.5–9	≤ 60	no	yes	< 15
MIT	++	(+)	(+)	slow – medium	2–10	≤ 80	no	no	100–200
BIT	(+)	(+)	(+)	slow	2.5–14	≥ 100	no	no	100–500
FA-Releaser	++	(+)	(+)	rapid	3–10	≤ 60	yes	(2)	100–1000
Bronopol	++	–	–	slow – medium	3–8.5	≤ 45	no	yes	50–300
DBDCB	+	+	+	slow	2.5–9	≤ 60	no	yes	250–1000
DBNPA	++	++	++	rapid	3–7[(3)]	≤ 70	no	yes	5–50

++ = higly effective; + = effective; (+) = effective, but with gaps; - = little effective

(1) approximate values (depending on the matrix properties)

(2) depending on the carrier molecule involved

(3) relatively stable in acidic aqueous media in the presence of stabilizers, rapid hydrolysis in alkaline media

Table 2.4: Selection of active combinations based on consumer requirements

Requirement(s)	Possible active combination
Best cost-performance ratio required	CMIT/MIT (3:1) + EDDM (O-formal)
Fast mode of action and head space protection required	CMIT/MIT (3:1) + O-formal CMIT/MIT (3:1) + N-formal
Fast mode of action required and no formaldehyde accepted	CMIT/MIT (3:1) + BIT CMIT/MIT (3:1) + Bronopol* CMIT/MIT (3:1) + DBDCB
No formaldehyde, no CMIT and no halogen in general accepted	BIT + MIT BIT + ZnP
No formaldehyde, no CMIT/MIT and no bronopol accepted	BIT + ZnP DBDCB + ZnP
Long lasting performance and head space protection required	BIT + O-formal BIT + N-formal
Long lasting performance required and no formaldehyde accepted	BIT + Bronopol* BIT + DBDCB BIT + MIT + Bronopol* BIT + MIT + DBDCB BIT + ZnP

* only very low amounts of formaldehyde to be expected under the usual processing conditions for coatings

2.1.3.3 2,2-Dibromo-3-nitrilopropionamide (DBNPA)

The active substance 2,2-dibromo-3-nitrilopropionamide (DBNPA; CAS No. 10222-01-2) (Figure 2.16) acts in a manner similar to the typical halogen biocides described above. It acts very rapidly and effectively by interfering with nucleophilic cell constituents, such as proteins in the cell membrane or enzymes. It provides broad-spectrum efficacy against gram-positive and gram-negative bacteria, fungi, yeasts, algae and is especially effective against slime-forming micro-organisms[127, 211]. DBNPA is therefore eminently suitable for slime control e.g. in paper making and in applications for inhibiting biofouling in cooling waters. Additionally, it serves as a preservative in pulp, slurries, soluble oil emulsions and in hydraulic fracturing, and numerous other areas. Due to rapid hydrolysis, especially in alkaline media, DBNPA is better suited to curative treatment, e.g. of infected raw materials or intermediates, rather than as a preservative for precautionary purposes[127, 211]. Its properties make DBNPA ideal for use as a sanitiser in the context of plant hygiene (see Section 2.3).

Thermal decomposition of DBNPA in formulations starts at approx. 70 °C. In aqueous alkaline media, it is rapidly hydrolyzed/degraded (see above). Nucleophilic agents, such as thiols, sulphites, bisulphites, hydrogen sulphide and other reducing agents can inactivate DBNPA, which is also susceptible to photodegradation (sunlight). Its solubility in water of approx. 1.5 % is not particularly marked but DBNPA is quite readily soluble in alcohols and in glycols[127]. The typical dosage varies with the intended purpose from 5 to 50 ppm.

2.1.4 Summary of relevant properties for in-can preservation

The following two charts (Table 2.3 and Table 2.4) offer guidance on whether a specific active substance or a combination of different actives should be selected for in-can protection purposes. Possible requirements in this context (see also Section 2.1), such as exclusion of specific chemistries by the user or compliance with a certain pH or temperature range in the product/system to be protected, render it essential to choose judiciously from the range of commercial microbicidal products available.

Some participants in the European coatings market use a very effective combination of 15 ppm CMIT/MIT (3 : 1), 150 to 200 ppm BIT and formaldehyde releaser with < 100 ppm free formaldehyde; this offers a good balance of performance requirements, commercial aspects and low risk of resistance formation by virtue of the fact that it employs four different microbicidal actives.

2.2 Dry-film preservation

Cured coatings undergo microbiological attack by fungi, algae and lichens, particularly in the presence of high atmospheric humidity (Figure 2.17). In the course of biocorrosion and biofouling processes, algae and fungi interfere by colonising, modifying and degrading these coatings, which often contain organic components, such as polymeric binders and additives that can act as a nutrient base for some of the microbial species involved[212].

In many European countries, architects and building companies are legally obliged to provide and erect a building that is free of defects for their clients, and this especially includes coatings that are free from microbial infestation – indoors and outdoors. To improve the protection of façades against growth of algae and fungi and therefore to prolong the service life of coating materials, the paints and coatings industry has developed water-repellent impregnation systems. However, these systems have not been as successful as expected as many of them are thermoplastic, a fact which attracts nutritive atmospheric particles to the uppermost surface layers, where local concentrations of moisture ultimately lead to microbial re-infestation by algae and fungi[9]. Consequently, it is still essential to use microbicides in building materials in order to extend their service life and to meet the growing demands of the construction industry. This is all the more important to consider, as the jury is still out

Figure 2.17: Examples of microbial infestation and deterioration of exterior surfaces

on the long-term efficacy of other coating developments, such as the "lotus effect", "low-energy", "self-cleaning" or "photocatalytic active nanoparticles", and there are some environmental concerns which need to be addressed as well[213–215].

Recent decades have seen a marked increase in microbial infestation of façades (see Figure 2.18 for examples). There are many reasons for this, including[9]:

- The environmental conditions conducive to microorganism growth have improved considerably owing to continuing eutrophication of the atmosphere with organic pollutants as well as to dramatic changes in the global climate.
- More-widespread application of external insulation systems on buildings as part of energy-saving measures. Thermal decoupling from the interior can lead to increased levels of condensation moisture on exterior façade surfaces, encouraging more microorganisms to settle and grow there.
- Architectural trends (e.g. flat roofs without overhang, leading to more exposure of the façade to driving rain)
- Poor maintenance and care due to economic reasons.

As long as microbicides remain efficacious against the microorganisms concerned, they help to control algal and fungal infestations of façades. However, their efficacy decreases over time due to consumption of the actives or natural decomposition *in situ*. In addition, climatic effects can cause fungicidal and algicidal actives to be leached out by rain or degraded by UV light.

With a view to improving the durability of its products, the paints and coatings industry is actively seeking future-proof solutions for keeping paint surfaces clean and free of infestation. Particularly when it comes to algicidal action, there is a very limited range of suitable microbicides available. Yet, there is growing demand for environmentally compatible, cost-efficient algicides, not least due to global climate change.

Although many factors govern the time response for microbial infestation of façades, the following are highly relevant[215]:

- Exposure conditions
- Execution of construction work
- Proper preparation and the quality of the applied coating system

Where microbial infestation occurs on a surface upon long-term exposure, indoors as well as outdoors, this is not necessarily and solely due to the stability of the microbicidal product in question or the outcome of poor leaching resistance. If, for instance, building deficiencies

Figure 2.18: Two examples of fungal and algal infestation of façade coatings

are involved, such as deficient thermal insulation (–› thermal bridges), then most preservation measures with microbicides are very likely doomed to fail in the long term unless such deficiencies are eliminated beforehand.

The intensified use of so-called external thermal insulation composite systems (ETICS), also known as exterior insulation and finishing systems (EIFS) (Figure 2.19), also has a substantial impact on the growth conditions of fungi and algae. Thermal decoupling of the exterior surface from the interior lowers the temperatures on external façades and can lead to increased condensation in those regions. As a result, fungi and algae find better growing conditions to thrive on. Furthermore, fungal growth is supported by the organic components in façade coatings.

In the field of interior coatings, thermal bridges caused by, say, poor workmanship during the installation of energy-saving measures can boost fungal growth inside a building. Inadequate isolation and construction faults can lead to remarkable temperature differences on interior wall sections (Figure 2.20). Hence, the temperature can partially fall below the dewpoint in those areas, giving rise to more condensation on the affected interior surfaces and, as a consequence, to a higher risk of fungal infestation.

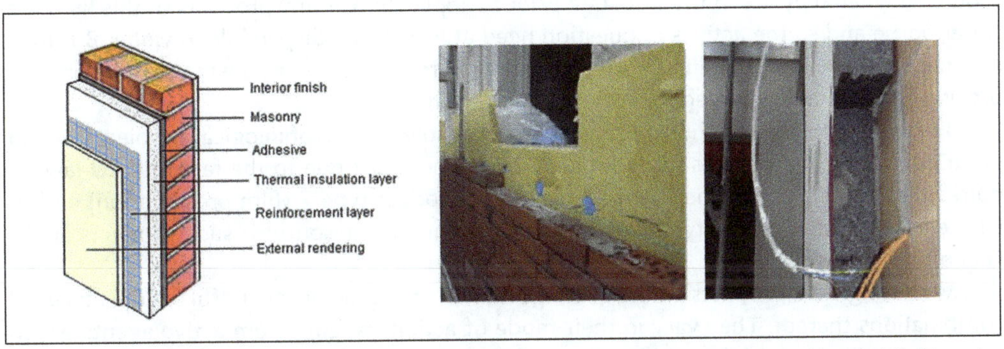

Figure 2.19: Schematic design of an 'external thermal insulation composite system' (ETICS) (left) and examples of thermal insulation measures in practice (centre and right)

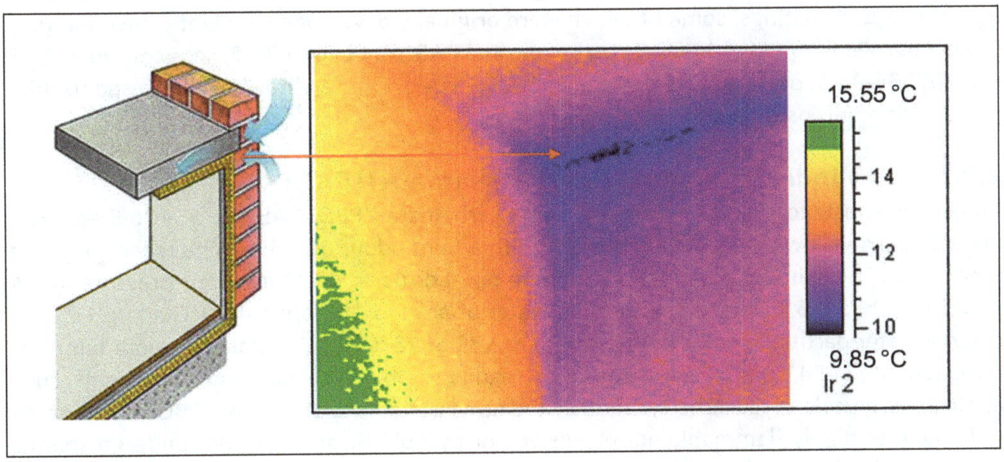

Figure 2.20: Risk of fungal infestation due to temperature differences on interior wall sections as visualised by infrared photography

Another aspect is the fact that state-of-the-art antimicrobial protection measures can contribute to sustainable product optimisation. The 'three-pillar model'[216] is predicated on the idea that sustainable development is only achievable if environmental, social and economic aims are pursued at the same time and are given equal importance.
Given optimum preservation of coatings:

- The substrate remains presentable over the long term and the integrity of the film matrix is retained for longer
- Expensive and resource-consuming redecoration is needed less often
- Risks to health can be minimised, e.g. in indoor wet areas
- The service life of coatings can be substantially extended and so natural resources can be conserved

2.2.1 Fungicides for coatings protection

Unlike the case for in-can microbicides, the actives present in film preservatives must have low solubility in water if they are not to be leached out by precipitation (e.g. rain). The other side of the coin is that suitable active substances need to be formulated in the products to be protected when they are in the wet state prior to application. Many products of this kind are water-borne and so the actives in question need at least to be dispersible in water. But they also need to have low volatility in order that the microbicidal active substances may be kept precisely where they are needed: in the surface layer.

The number of active substances available for solving microbiological problems in and on dried coating materials is fairly limited. This is not least due to the fact that EU law requires that a complete dossier be filed for use in product type 7 (film preservation) and/or where appropriate product type 10 (protective agents for masonry) if such products are to stay on the market.

Most fungicidal products on the coatings market are based on a handful of substances or combinations thereof. They vary in their mode of action, ranging from active agents which are electrophilic, such as the isothiazolinones already mentioned in Section 2.1.2, substances which interfere with cell membranes and finally systemic fungicides which interact with diverse cell functions (see also Table 1.3). Figure 2.21 provides an overview of fungicides frequently used in coatings, some of which were originally developed for plant protection (e.g. thiabendazole, CAS No. 148-79-8; carbendazim, CAS No. 10605-21-7, tebuconazole, CAS No. 107534-96-3; propiconazole, CAS No. 60207-90-1). They are now being utilised for film preservation purposes as well.

2.2.1.1 3-Iodo-2-propynyl butyl carbamate (IPBC)

The active substance 3-iodo-2-propynyl butyl carbamate (IPBC, CAS No. 55406-53-6) (Figure 2.21) is a yellowish crystalline powder with a faint odour of iodine. The technical grade (approx. 98 % minimum purity) has a melting point of 65 to 67 °C, a vapour pressure of 2.4 to 4.5 x 10^{-3} Pa (25 °C) and a solubility in water of around 168 ppm at 20 °C and pH 7. The octanol/water partition coefficient (log P_{OW}) is 2.81 at 25 °C. IPBC is stable at room temperature (stability at 54 °C: ≥ 14 days) and fairly readily soluble in certain organic solvents, such as alcohols, glycols, glycol ethers, acetone, and ethyl acetate. Furthermore, IPBC is considered to be neither highly flammable nor explosive nor pyrophoric and does not undergo spontaneous combustion[217, 218]. In some cases, it can induce yellowing in the coating system to be protected.

The Cosmetic Products Regulation[185] authorises IPBC concentrations of below 0.02 % (200 ppm) in rinse-off products, below 0.01 % (100 ppm) in leave-on products and below 0.0075 % (75 ppm) in deodourants/anti-perspirants.

In the context of European biocides legislation, applications have been submitted under Directive 98/8/EC[181] to approve IPBC for use in the following product types: 6 (preservatives for products during storage), 7 (film preservatives), 8 (wood preservatives), 9 (fibre, leather, rubber and polymerised materials preservatives), 10 (construction material preservatives) and 13 (working or cutting fluid preservatives). The active substance already has BPD and BPR approval for use in PTs 6, 8, and 13[217].

When the European authorities conducted a human health risk assessment (HHRA) on IPBC[218], it was found to be neither carcinogenic nor neurotoxic nor genotoxic. Furthermore, it was stated that the substance was not toxic to reproduction and not a developmental toxicant. The active was also described as non-irritating to skin but as eye irritant and a skin sensitiser. IPBC was reported to possess moderate acute toxicity by the oral route and low toxicity by the dermal route. Additionally, IPBC is classified as toxic by inhalation. Histopathological findings for inhalation (repeated exposure) of solid IPBC indicated an effect in the central region of the larynx. However, this result was regarded as being local rather than systemic.

The environmental risk assessment (ERA) indicated *inter alia* primary biodegradability in the Zahn-Wellens test (degrades rapidly within 2 hours to propargyl butyl carbamate).

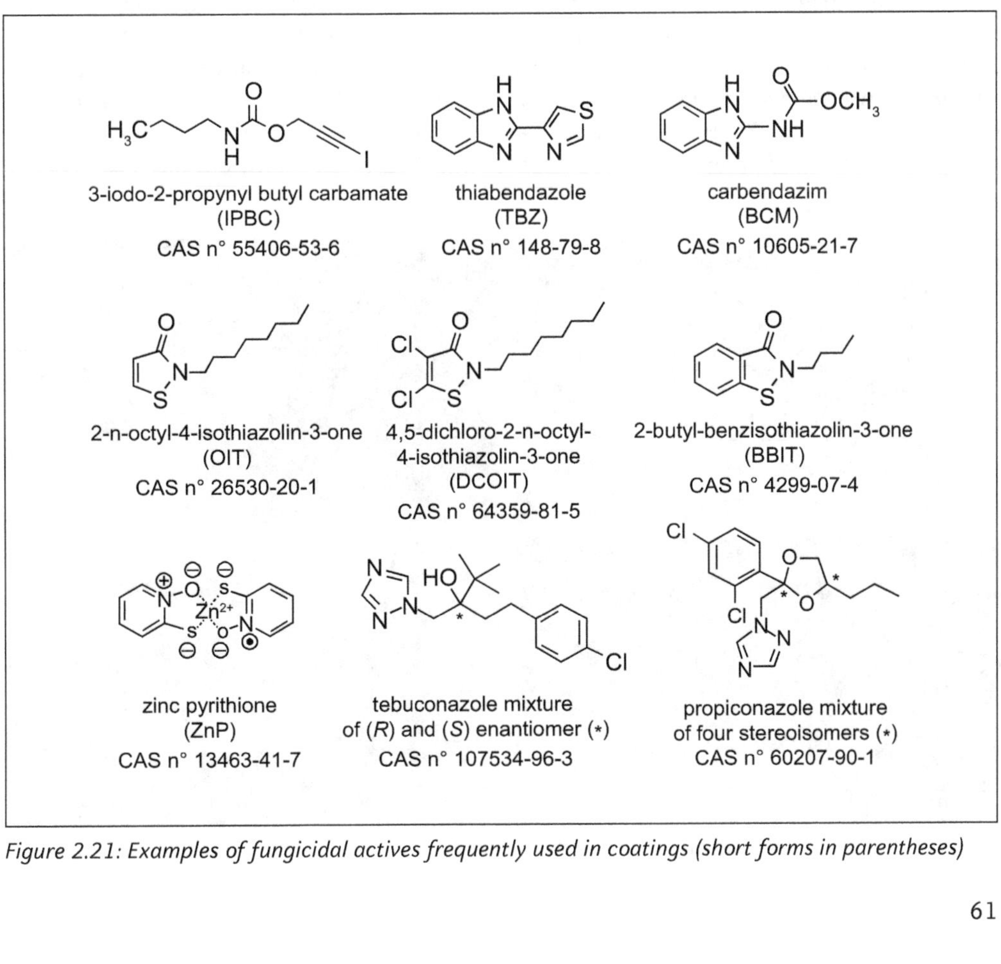

Figure 2.21: Examples of fungicidal actives frequently used in coatings (short forms in parentheses)

A great many more details related to human health hazards and environmental impact can be found in the respective Assessment Reports by the Member State Competent Authorities[217, 218].

IPBC disturbs cell membrane permeability in fungal cells and affects the fatty acid metabolism[129] (see also Table 1.3). It offers broad efficacy against a wide variety of fungi relevant to coatings, including *Alternaria spp.* It is also active against a couple of bacteria and several algal species. Additionally, IPBC is highly effective against blue stain fungi (e.g. *Aureobasidium pullulans*) and possesses a certain degree of efficacy against some wood-rotting fungi. Consequently, it is also employed in wood protection (product type 8), especially as a fungicidal combination partner for azole derivatives, such as tebuconazole and propiconazole (dedicated actives for combating wood-destroying fungi) and optionally together with insecticides/termiticides, e.g. pyrethroid derivatives (see also Table 1.3).

The term 'blue stain' describes a discolouration effect which is generated by certain fungi on wood and which manifests itself as a blue to greyish-black surface stain (Figure 2.22). This stain makes the wood unsuitable for some applications, although the material itself is not destroyed by these microorganisms. The typical dosage in coatings is 400 to 2000 ppm. For wood protection, higher dosages of 3000 to 10,000 ppm IPBC are not unusual.

2.2.1.2 Thiabendazole (TBZ)

The active substance thiabendazole (IUPAC name: 2-thiazol-4-yl-1H-benzoimidazole, CAS No. 148-79-8) (Figure 2.21) is a pale beige, odourless powder at room temperature. The technical grade (approx. 98.5 % purity) has a melting point of 297 to 298 °C, a low vapour pressure of 4.6 to 5.3 x 10^{-7} Pa (25 °C) and a solubility in water of approx. 30 ppm at 20 °C and pH 7, with two dissociation constants (pKa: 4.73 and 12.00). Its solubility in organic solvents is quite limited. The Henry's law constant indicates that significant volatilisation in the environment is not to be expected. The octanol/water partition coefficient (log P_{OW}) is 2.39

Figure 2.22: Examples of infestation of wood substrates by blue stain and wood-rotting fungi

at 20 °C and pH 7. Thiabendazole is stable at room temperature and thermally decomposes at about 310 °C. Furthermore, TBZ is neither highly flammable nor explosive nor pyrophoric and is non-oxidising. It does not ignite below its melting point and nor does it evolve any flammable gases in contact with water or humid air[219, 220].

In connection with the European biocides legislation, applications that have been submitted under Directive 98/8/EC[181] to approve TBZ for use in the following product types are being assessed: 7 (film preservatives), 8 (wood preservatives), 9 (fibre, leather, rubber and polymerised materials preservatives) and 10 (construction material preservatives). The active substance already has BPD approval for use in PT 8[217].

Evaluation by the European authorities[219, 220] indicates that thiabendazole is not irritating to eyes and skin and that the active is not a dermal sensitiser in guinea-pigs. TBZ does not exhibit any genotoxic potential *in vitro* or *in vivo* nor give rise to any reproductive toxic effects at parental toxic doses in rats. Additionally, the substance is reported as being non-teratogenic. For further details, the reader is referred to the reports compiled by the respective authorities[219, 220].

Unlike the majority of film preservatives employed in that field, thiabendazole is not classified as a 'human health hazard' under CLP Regulation[196]. This makes thiabendazole a highly attractive candidate for use as film preservative in specialised indoor applications, especially for combating fungal contamination confined to localised wet areas, such as bathrooms and laundry compartments.

Thiabendazole is described as a contact, systemic benzimidazole fungicide that has protective properties and a curative action. The substance acts directly on and interferes with the assembly of microtubules (components of the fungal exoskeleton) by binding to their protein sub-unit beta-tubulin[154–156, 219].

Apart from providing film protection in coatings in a typical dosage of 1000 to 4000 ppm active substance, TBZ is approved under European plant protection legislation[314] and is used, for example, to combat vegetable and fruit diseases (mould, rot, blight, and stain) and for controlling storage diseases. Additionally, it can be used to treat several helminthic species (e.g. roundworms) in livestock and humans and medicinally as a chelating agent to bind metals[221].

TBZ is effective against mould and blue stain fungi, but there are major gaps in its effectiveness at combating *Alternaria spp.* which frequently occurs on exterior surfaces. Additionally, it has limited efficacy against yeasts and is not bactericidal or algicidal. However, thiabendazole offers excellent fungicidal performance in combination with other actives that can close these gaps, such as OIT, IPBC, and zinc pyrithione (see below), and is therefore an interesting combination partner in fungicidal products for material protection.

2.2.1.3 Carbendazim (BCM)

The active substance carbendazim (IUPAC name: methyl benzimidazol-2-ylcarbamate, CAS No. 10605-21-7) (Figure 2.21) is a sandy to light-grey crystalline, odourless powder. The technical grade (approx. 98 % minimum purity) has a melting point of 302 to 307 °C (with decomposition), a vapour pressure of 1.5×10^{-4} Pa (25 °C) and a solubility in water of 5 to 7 ppm (pH 7 to 8, ambient temperature). The octanol/water partition coefficient (log P_{OW}) is 1.50 at pH 5 to 9. Carbendazim is stable at room temperature, and is considered to be neither highly flammable nor explosive[222, 223].

With regard to European biocides legislation, applications have been submitted to approve carbendazim for use in product types 7 (film preservatives), 9 (fibre, leather, rubber and polymerised materials preservatives) and 10 (construction material preservatives)[217].

The active was formerly approved under European plant protection legislation, but approval was not extended after expiration in 2014[314].

Carbendazim is reported as being not acutely toxic via the oral, dermal or inhalation routes. It is not considered to be a skin or eye irritant, but it is indicated as a skin sensitiser. Furthermore, there is evidence that the active substance is mutagenic (both *in vitro* and *in vivo*) and toxic to reproduction and development[222]. Consequently, BCM is currently classified as "Muta. 1B, H340: May cause genetic defects" and as "Repr. 1B, H360FD: May damage fertility. May damage the unborn child" under the CLP Regulation[196]. Thus, carbendazim meets two of the exclusion criteria of the BPR (see Section 5)[126].

Carbendazim belongs to the group of substances which are prohibited under the Cosmetic Products Regulation[185]. Further details on the hazard properties can be found in review reports from European authorities[222, 223].

In fungi, the active substance is reported to interfere with DNA synthesis and related processes, such as cell division. Neither synthesis of RNA nor protein synthesis appears to be affected[224]. Another reference indicates interference of carbendazim with mitosis, cytokinesis and DNA synthesis in two fungal species[225].

BCM is efficacious against a variety of fungi and yeasts, including wood-staining fungi, but, like thiabendazole, it has a major gap in respect of *Alternaria spp.*, which usually occurs outdoors. The microbicidal action of carbendazim on bacteria and algae is poor. Its low solubility in water and in organic solvents makes it difficult to formulate. When used in combination with other actives that can close the gaps of carbendazim, such as OIT, IPBC or zinc pyrithione, complete protection against fungal infestation can be achieved. Carbendazim is typically dosed at below 1000 ppm, this figure being the limit value that would trigger a classification and labelling of the product to be protected.

2.2.1.4 Zinc pyrithione (ZnP)

The active substance zinc pyrithione (IUPAC name: zinc bis(2-thioxopyridin-1(2H)-olate), CAS No. 13463-41-7) (Figure 2.21) is a dull-white, odourless powder. The technical grade (98 to 99 % purity) has a melting point of > 240 °C (with decomposition), a vapour pressure less than 1×10^{-6} Pa (25 °C) and a solubility in water of 5 to 6 ppm at 20 °C. The octanol/water partition coefficient (log P_{OW}) is 0.88 at pH 6.4 and 20 °C. ZnP is sparingly soluble in many organic solvents. The chelated zinc complex is stable in the pH range of 4 to 8.5. In acidic media (< pH 3.5) the active is converted to pyrithione while in alkaline media (> pH 8.5) conversion to water-soluble alkali salts can be detected. Zinc pyrithione is also susceptible to strong oxidising and reducing agents as well as to light exposure. The presence of heavy metals (especially iron, copper and cobalt), even in very low quantities, may lead to the formation of trans-chelation products which may give rise to significant colouration. The active is considered to be neither highly flammable nor explosive nor oxidising and no relevant self-ignition temperature exists below its melting point[127, 226–228]. Diverging values for some physicochemical properties are reported in different sources of information.

Zinc pyrithione is authorised under the Cosmetic Products Regulation[185] as a preservative at a maximum concentration of 1 % in rinse-off hair products and 0.5 % in other rinse-off products. In leave-on hair products, the allowable maximum concentration is 0.1 % for purposes other than inhibiting the development of microorganisms in the product. Use in oral hygiene products is forbidden.

In the context of European biocides legislation, applications to approve zinc pyrithione for use in the following product types are being assessed: 2 (disinfectants and algicides not intended for direct application to humans and animals), 6 (preservatives for products during storage), 7 (film preservatives), 9 (fibre, leather, rubber and polymerised materials preservatives), 10 (construction materials preservatives) and 21 (antifouling products)[217].

Registrant(s) under REACH have suggested classification for eye damage, toxicity if swallowed and if inhaled,[226, 227]. Moreover, zinc pyrithione is described as exhibiting no sensitising effects in animals. However, the Member State Competent Authority Sweden submitted in October 2016 a proposal for harmonised classification and labelling (CLH dossier), *inter alia* with the proposed classification as being reprotoxic 1B (H360D; may damage the unborn child)[228]. With this, zinc pyrithione would meet the exclusion criteria of the BPR.

Zinc pyrithione belongs to the class of membrane-active substances which have chelating properties that affect ATP levels, protein synthesis and nutrient transport[147–150].

It is active against a wide variety of bacteria, moulds, yeasts, and a couple of algal species. ZnP is therefore frequently used for dry-film preservation of architectural paints, especially in combination with other fungicides, such as OIT, BCM, IPBC and TBZ. The typical dosage of this active for dry-film preservation is 500 to 2000 ppm.

Extender pigments, such as calcium carbonate, often contain small amounts of iron compounds which may form deeply coloured trans-chelation products in combination with zinc pyrithione. Addition of approx. 500 ppm zinc oxide (expressed in terms of wet paint) can protect the formulations in question from these discolouration effects. However, such effects also occur in the presence of cobalt ions and are not easily avoided. As a result, ZnP is usually not employed in those alkyd paints which contain drier catalysts based on cobalt compounds[229].

On account of its bactericidal properties, zinc pyrithione is also used for in-can protection, e.g. in combination with the active BIT (e.g. BIT/ZnP in the ratio 1 : 1 or 1.5 : 1; see also Section 2.1.2), for protecting water-borne formulations, such as paints in the wet state. The dosage of zinc pyrithione in this case is 14 to 200 ppm.

Zinc pyrithione is also employed in antifouling marine paints at a dosage of 1 to 5 %, where it acts as a co-biocide to combat soft-fouling (e.g. fungi, algae, diatomaceous slimes) and it is usually combined with cuprous oxide (35 to 40 %), which protects the marine paint from hard-fouling (barnacles, tubeworms, mussels etc.).

2.2.1.5 2-n-octyl-4-isothiazolin-3-one (OIT)

The active substance OIT (IUPAC name: 2-octyl-1,2-thiazol-3(2H)-one, CAS No. 26530-20-1) (Figure 2.21) is a yellow to amber-coloured liquid with a mild odour. The technical grade (approx. 99 % purity) has a boiling point of 120 °C at 0.013 hPa, an octanol/water partition coefficient (log P_{OW}) of 3.42, a vapour pressure of 4.9 x 10^{-3} Pa (25 °C) and a solubility in water of approx. 525 ppm at ambient temperature. OIT is quite readily soluble in diverse organic solvents. It is sensitive to oxidising and reducing agents as well as to attack by nucleophiles[230, 231].

In connection with the European biocides legislation, applications for the approval of OIT for use in the following product types are being assessed: 6 (preservatives for products during storage), 7 (film preservatives), 8 (wood preservatives), 9 (fibre, leather, rubber and polymerised materials preservatives), 10 (construction materials preservatives), 11 (preservatives for liquid-cooling and processing systems) and 13 (working or cutting fluid preservatives)[217].

ECHA published in December 2016 its opinion on the application for approval of the active substance OIT for PT 8 in which a new classification according to the CLP Regulation is proposed[232]. The proposed future entry in Annex VI of CLP Regulation[196] is classification for toxicity if swallowed, in contact with skin and by the inhalation route. The substance is further indicated to be corrosive to the respiratory tract and to the skin and classification as skin sensitiser (1A, H317) with a specific concentration limit (SCL) of 50 ppm has been proposed. Consequently, the EUH Phrase 208 ('contains 2-octyl-1,2-thiazol-3(2H)-one. May produce an allergic reaction') would apply to all formulations containing OIT in the range of 5 to < 50 ppm, provided that this proposal be adopted under the CLP Regulation.

OIT belongs to those electrophilic active agents with activated N-S bond which react with nucleophiles, such as amines and thiols. In particular, the reaction with intracellular thiols of biocatalysts leads to enzyme inhibition within minutes. The destruction of protein thiols and the generation of free-radicals finally induce cell death within hours[158, 159].

OIT offers excellent efficacy against a wide range of fungi, but it also covers diverse bacterial species. In coatings, it mainly serves as a film fungicide. The typical dosage for this application is 250 to 500 ppm, the latter being the current specific concentration limit for classification and labelling as skin sensitiser for products preserved with OIT. One weakness of this active is its low resistance to leaching. Consequently, exterior coatings containing only OIT as fungicide are possibly at risk of microbial infestation after a period of outdoor exposure. OIT is often combined with other fungicides, such as BCM, IPBC, TBZ, and ZnP, for film preservation purposes.

2.2.1.6 4,5-Dichloro-2-n-octyl-4-isothiazolin-3-one (DCOIT)

The active substance DCOIT (IUPAC name: 4,5-dichloro-2-octylisothiazol-3(2H)-one, CAS No. 64359-81-5) (Figure 2.21) is an off-white, odourless solid. The relevant assessment reports by European bodies[233, 234] stipulate that the minimum purity of the active substance as manufactured must lie within the range 95 to 100 %. However, some values relating to physicochemical properties are provided for technical grade of 98 to 99 % purity. These values are also stated here.

DCOIT has a melting point of 41 to 42 °C, an octanol/water partition coefficient (log P_{OW}) of 2.8 at pH 7 and 23 °C, a vapour pressure of 9.8 x 10^{-4} Pa at 25 °C and a solubility in water of approx. 3.5 ppm at pH 7 and 20 °C. DCOIT is quite readily soluble in diverse organic solvents, such as ethanol, acetone, toluene, ethyl acetate, and hexane. Furthermore, DCOIT is considered to be neither highly flammable nor auto-flammable, nor explosive. The substance is sensitive to oxidising and reducing agents as well as to nucleophilic attack, e.g. by amines. Its stability in the wet state might be limited in some cases[233, 234].

With regard to European biocides legislation, applications have been submitted to approve DCOIT for use in the following product types: 7 (film preservatives), 8 (wood preservatives), 9 (fibre, leather, rubber and polymerised materials preservatives), 10 (construction material preservatives), 11 (preservatives for liquid-cooling and processing systems) and 21 (antifouling products). The active substance already has BPD and BPR approval for use in PTs 8 and 21[217].

As part of the harmonised classification and labelling process (CLH) throughout the European Union, a dossier was submitted by the Norwegian Competent Authority[235]. The proposed future entry in Annex VI of CLP Regulation[196] is for classification with Acute Tox. 4 for the oral route (H302: Harmful if swallowed) and Acute Tox. 1 for the inhalation route (H330: Fatal if inhaled). The substance is further indicated to cause severe skin burns and eye damage, and classification as skin sensitiser (1A, H317) with a specific concentration limit (SCL) of 10 ppm is proposed. This means that the phrase EUH 208 ('contains 4,5-dichloro-2-octyl-isothiazol-3(2H)-one. May produce an allergic reaction') would apply to all formulations containing DCOIT in the range 1 to < 10 ppm, provided that this proposal is adopted in the CLP

Regulation. The proposed specific concentration limit for skin irritation is 100 ppm. As a consequence, doses of 10 to < 100 ppm DCOIT would trigger the respective classification and labelling of the formulations to be protected (e.g. a paint) as skin sensitising and, at dosages of 100 ppm or higher, as skin irritating.

There is no evidence of mutagenicity, carcinogenicity, reproductive toxicity or neurotoxicity according to the information in the assessment reports[233, 234].

DCOIT, like OIT, is an electrophilic active agent and contains an activated N-S bond. However, as a chlorinated isothiazolinone derivative, it additionally possesses vinyl-activated halogens. The presence of two toxophoric structural elements in a single molecule improves the antimicrobial efficacy relative to the corresponding non-halogenated variant[189, 190]. DCOIT reacts with the proteins of the microorganisms, interrupting important metabolic processes that utilise these proteins[233, 234].

DCOIT provides broad antimicrobial activity against fungi, yeasts, bacteria, algae and sea organisms at very low inhibitory concentrations. In coatings, the typical dosage is 100 to 1000 ppm.

DCOIT is also employed as a broad-spectrum co-biocide in a concentration of 1 to 3 % in antifouling paints for controlling soft-fouling (e.g. fungi, algae, diatomaceous slimes) and is usually combined with cuprous oxide, which protects marine paint from hard-fouling (barnacles, tubeworms, mussels etc.).

2.2.1.7 2-n-Butyl-1,2-benzisothiazolin-3-one (BBIT)

The active substance BBIT (IUPAC name: 2-butyl-1,2-benzothiazol-3(2H)-one, CAS No. 4299-07-4) (Figure 2.21) is a brown liquid with no obvious odour. The technical grade (95.5 to 97 % purity) has an octanol/water partition coefficient (log P_{OW}) of 2.86 at 25 °C and pH 6.7 to 6.9, a vapour pressure of 1.5 x 10^{-2} Pa (25 °C) and a solubility in water of 760 to 800 ppm (10 to 30 °C, pH 7). BBIT is readily soluble in diverse organic solvents. Thermal decomposition starts at approx. 300 °C. It is considered to be neither explosive nor oxidising nor self-igniting. It is stable over a wide pH range (pH 2 to 12) but is sensitive to attack by nucleophiles, such as thiols[127, 236].

In the context of European biocides legislation, applications to approve BBIT for use in the following product types are being assessed: 6 (preservatives for products during storage), 7 (film preservatives), 9 (fibre, leather, rubber and polymerised materials preservatives), 10 (construction material preservatives) and 13 (working or cutting fluid preservatives)[217].

According to the current entry in Annex VI of the CLP Regulation[196], the active substance is classified as skin corrosive (1B, H314) and skin sensitising (1, H317), both without specific concentration limits. The phrase EUH 208 ('contains 2-butyl-1,2-benzothiazol-3(2H)-one. May produce an allergic reaction') would apply to all formulations containing BBIT in the range 1000 to < 10000 ppm.

BBIT was not mutagenic in gene-mutation tests (*in vivo* micronucleus test was negative but there were positive results in an *in vitro* chromosome aberration test) and the active is not considered to be carcinogenic[236].

The mode of action of BBIT is expected to be similar to that of 1,2-benzisothiazol-3(2H)-one (BIT). The latter is reported to affect thiol-dependent enzymes which play an important role in transport processes across the cytoplasmic membrane and to impact the electron transport chain utilised by bacteria[236].

BBIT offers a broad efficacy spectrum against fungi, yeasts, a couple of algal species and a range of bacteria, but is only moderately effective against *Pseudomonas spp.* Besides its use as film preservative in coatings at a typical dosage of 1000 to 3000 ppm, the active is also used to provide wet-state protection (dosage: 250 to 500 ppm)[236].

The efficacy of BBIT over a wide pH range (pH 2 to 12), together with its chemical and thermal stability, renders it ideal for the protection of plastics and metalworking fluids.

2.2.1.8 Tebuconazole

The active substance (IUPAC name: (RS)-1-(4-chlorophenyl)-4,4-dimethyl-3-(1H-1,2,4-triazol-1-ylmethyl)-pentan-3-ol, ratio (1 : 1), CAS No. 107534-96-3) (Figure 2.21) is a white to beige powder with a slight characteristic odour. The purity of the technical grade is specified at ≥ 95 % of a racemic mixture (R- and S- enantiomer) in the ratio 1 : 1. Tebuconazole has a melting point of 105 °C (99.9 % purity), an octanol/water partition coefficient (log P_{OW}) of 3.49 at 20 °C, a vapour pressure of 1.7×10^{-6} Pa (20 °C) and a solubility in water of 29 ppm (pH 7, 20 °C). It is readily soluble in diverse organic solvents, such as aliphatic alcohols, acetone, toluene, and acetonitrile. Tebuconazole is considered to be neither highly flammable nor pyrophoric nor explosive and it does not exhibit oxidising properties. Contact with water or humid air does not lead to the evolution of any flammable gases and there is no self-ignition at temperatures up to the melting point[237, 238].

In connection with the European biocide legislation, applications to approve tebuconazole for use in the following product types are being assessed: 7 (film preservatives), 8 (wood preservatives) and 10 (construction material preservatives). The active substance has already been approved under BPD and BPR for all product types[217]. In addition, tebuconazole is approved under European plant protection legislation[314].

According to the current entry in Annex VI of the CLP Regulation[196], the active substance is classified as toxic to reproduction (Repr.2, H361d: Suspected of damaging the unborn child) and rated Acute Tox. 4 by the oral route (H302: harmful if swallowed). Tebuconazole has already been assessed by European authorities[237, 238] as being of low toxicity by inhalation and by the dermal route. Additionally, it exhibits low to moderate toxicity by the oral route. There is no evidence of skin- or eye-irritation potential or of skin-sensitising properties. Furthermore, genotoxic potential has not been observed in *in vitro* and *in vivo* tests and nor is there any evidence of carcinogenicity. In developmental toxicity studies, foetotoxic effects were observed in all species under test. Accordingly, tebuconazole is classified as being toxic to reproduction (see before).

Tebuconazole is considered a candidate for substitution[239] under Article 10(1)(d) of the BPR[126] as it meets two of the criteria for being PBT (persistent, bioaccumulative and toxic) under Annex XIII to Regulation (EC) No. 1907/2006.

Tebuconazole inhibits the biosynthesis of ergosterol in fungal membranes, which at low azole concentrations leads to retardation of fungal growth and at high azole concentrations to disruption of fungal membranes[132, 135, 151–153].

Tebuconazole is a declared fungicide with high efficacy against wood-destroying fungi (Basidiomycetes), although there are gaps in its efficacy against blue-staining fungi and some other species, such as *Alternaria spp.* and *Trichoderma spp.* It is frequently employed in wood preservation (PT 8), particularly in combination with IPBC, as a highly effective microbicide for combating blue stain fungi. For wood-protection purposes, typical biocidal products (both water-borne and solvent-borne) are described as containing up to 6000 ppm tebuconazole[240]. The coverage of such products lies in the range 120 to 180 ml/m² wood substrate. In film preservation (PT 7), typical end-use products are quoted as being used indoors to seal joints against ingress of moisture (e.g. in shower cabins, bathtubs, kitchen sinks etc.). In this case, a dosage up to 5200 ppm of the active substance is indicated for preventing fungal infestation[238]. The same dosage level is described in outdoor applications, where tebuconazole serves as a masonry preservative (PT 10) for sealing façade joints[241].

2.2.1.9 Propiconazole

The active substance propiconazole (IUPAC name: (2RS,4RS;2RS,4SR)-1-[[2-(2,4-dichlorophenyl)-4-propyl-1,3-dioxolan-2-yl]methyl]-1H-1,2,4-triazole, CAS No. 60207-90-1) (Figure 2.21) is, viscous, odourless, yellowish liquid. The purity of the technical grade is specified at ≥ 96 % for a mixture of four stereoisomers that are all biocidal. Propiconazole has a boiling point of > 250 °C (92.2 % purity), an octanol/water partition coefficient (log P_{OW}) of 3.72 at pH 6.6 and 25 °C (99.1 % purity), a vapour pressure of 5.6 x 10^{-5} Pa (25 °C, 99.1 % purity) and a solubility in water of approx. 100 ppm (pH 6.9, 20 °C, 99.1 % purity). It is readily soluble in diverse organic solvents, such as toluene, dichloromethane, ethanol, n-octanol, acetone, and ethyl acetate. The active substance at a purity of 92.4 % is considered to be neither flammable nor explosive nor oxidising and it does not undergo self-ignition before it starts decomposing at 355 °C[242].

With regard to European biocide legislation, applications for the approval of propiconazole for use in the following product types are being assessed: 7 (film preservatives), 8 (wood preservatives) and 9 (fibre, leather, rubber and polymerised materials preservatives). The active substance already has BPD and BPR approval for all applied product types[217]. Propiconazole is also approved under European plant protection legislation[314].

Within the framework of the harmonised classification and labelling process (CLH) throughout the European Union, a dossier has been submitted by the Finnish Competent Authority. The here proposed future entry in Annex VI of CLP Regulation[196] provides for classification under reproduction toxicity (Repr.2, H361d: Suspected of damaging the unborn child), with Acute Tox. 4 (H302: harmful if swallowed) and with Skin Sens.1 (H317: May cause an allergic skin reaction). However, the responsible Committee for Risk Assessment (RAC) which evaluates the CLH dossiers came to the conclusion that a more severe classification for reproduction toxicity under category 1B is justified[243]. Consequently, the active substance would fall under the exclusion criteria under BPR provided that this proposal be adopted under the CLP Regulation.

Propiconazole has been evaluated by European bodies[242] as being moderately toxic by the oral route and to be a moderate skin sensitiser. However, there is no evidence of skin corrosion/irritation properties or of eye irritation. Effects reported in animal studies were related to carcinogenicity and reproductive toxicity. Many more details related to human health hazards and environmental impact can be found in the Assessment Report by the Competent Authority[242].

Propiconazole inhibits the biosynthesis of ergosterol in fungal membranes, which leads to retardation of fungal growth at low azole concentrations and to disruption of fungal membranes at high azole concentrations[132, 135, 151–153].

The activity spectrum of propiconazole is similar to that of tebuconazole (see before). It is therefore also frequently employed in wood preservation (PT 8), quite often together with IPBC as a microbicidal combination partner that closes the gaps against blue stain fungi. For use in film preservation (PT 7), a dosage range of 1000 to 3000 ppm active substance for preventing fungal infestation on paint films is indicated in the Competent Authority Report[242]. A diverse range of microbicidal products intended for film preservation contain at least two different fungicidal actives for closing potential gaps in the corresponding combination part-

Table 2.5: Contrasting important properties of film fungicides – Part I

	IPBC	**TBZ**
CAS n°	55406-53-6	148-79-8
Physical state	yellowish, crystaline powder	pale beige powder
Odour	faint odour of Iodine	odourless
Technical grade a.i.	~ 98 %	~ 98.5 %
Melting point	~ 65–67 °C	~ 297–298 °C
Boiling point		
Vapour pressure [Pa]	2.4–4.5 x 10^{-3} (25 °C)	4.6–5.3 x 10^{-7} (25 °C)
water solubility [ppm]	168 (pH 7, 20 °C)	30 (pH 7, 20 °C)
log P_{OW}	2.81 (25 °C)	2.39 (pH 7, 20 °C)
Application for approval (PT) under EU biocide legislation	6, 7, 8, 9, 10, 13	7, 8, 9, 10
Approval (PT) under EU biocide legislation (as per May 2017)	6, 8, 13	8
Evaluating Competent Authority	Denmark	Spain
Chemisry	carbamate	benzimidazole
Mode of action	disturbs the cell membrane permeability and affects the fatty acids metabolism	contact and systemic benzimidazole fungicide; interferes with the assembly of microtubules (components of fungal exoskeleton)
Activiy	effective against a wide variety of fungi (including *Alternaria spp.* & blue stain fungi) as well as against bacteria and several algae species	protective and curative action; effective against mold and blue stain fungi, major gaps against *Alternaria spp.* and yeasts; no bactericidal and no algicidal activity
Approximate dosage in the coatings area [ppm]	400–2000	1000–4000
Remarks	authorized under the Cosmetic Products Regulation; yellowing effects possible	no human health hazard acc. to GHS CLP; approved under EU plant protection legislation

ner's activity spectrum, so that the above-mentioned dosage levels should usually deliver satisfactory results when used in conjunction with the proper combination products.

For wood protection purposes, typical commercial biocidal products (both water-borne and solvent-borne) contain 6000 to 10,000 ppm propiconazole. The coverage of such products lies in the range 120 to 180 ml/m² wood substrate.

2.2.1.10 Summary of important properties of film fungicides

Table 2.5 and Table 2.6 contrast the important properties of the fungicidal actives described here which are frequently used in the coatings area for dry-film preservation purposes.

BCM	**ZnP**
10605-21-7	13463-41-7
sand-colored to light grey crystalline powder	dull white powder
odourless	odourless
~ 98 %	~ 98–99 %
~ 302–307 °C	≥ 240 °C
1.5×10^{-4} (25 °C)	$< 1 \times 10^{-6}$ (25 °C)
5–7 (pH 7–8, RT)	5–6 (20 °C)
1.50 (pH 5–9)	0.88 (pH 6.4, 20 °C)
7, 9, 10	2, 6, 7, 9, 10, 21
–	–
Germany	Sweden
benzimidazole; carbamate	pyridine-N-oxide
Interference with DNA synthesis, mitosis, cytokinesis or related processes	membrane-active substance with chelating properties; influence on ATP levels, protein synthesis and nutrient transport
efficacy against a variety of fungi and yeasts including wood staining fungi, major gap against *Alternaria spp.*; microbicidal action against bacteria and algae is poor	covers bacteria, molds and yeasts and also a couple of algae species; due to bactericidal properties also utilization as in-can preservative
< 1000	500–2000 (PT 7) 14–200 (PT 6)
meets two exclusion criteria under BPR (Muta. 1B, Repr. 1B); no longer approved under EU plant protection legislation	proposed for Repr. 1B classification (exclusion criterion under BPR); discoloration issues; sensitive to light

2.2.2 Algicides for coatings protection

Like film fungicides, algicides should have low solubility in water if they are to minimise potential leaching from the cured paint film. Here, again, suitable active substances usually need to be incorporated into biocidal products intended for film preservation in the wet state. This needs to be done before the final product is applied to the surface to be protected. Many such products are water-borne and so the relevant algicidal ingredients should at least be dispersible in water. On the other hand, their volatility should ideally be low to avoid evaporation of the microbicidal active from the cured paint film. If obtainable, the substances in question should be resistant to UV radiation or to oxidising and reducing agents. Such influ-

Table 2.6: Contrasting important properties of film fungicides – Part II

	OIT	**DCOIT**
CAS n°	26530-20-1	64359-81-5
Physical state	yellow to amber liquid	off-white solid
Odour	mild odour	odourless
Technical grade a.i.	~ 99 %	~ 98 – 99 %
Melting point		~ 41 – 42 °C
Boiling point	120 °C (0,013 hPa)	
Vapor pressure [Pa]	4.9×10^{-3} (25 °C)	9.8×10^{-4} (25 °C)
water solubility [ppm]	525 (RT)	3.5 (pH 7, 20 °C)
log P_{OW}	3.42	2.8 (pH 7, 23 °C)
Application for approval (PT) under EU biocide legislation	6, 7, 8, 9, 10, 11, 13	7, 8, 9, 10, 11, 21
Approval (PT) under EU biocide legislation (as per May 2017)	–	8, 21
Evaluating Competent Authority	England	Norway
Chemisry	isothiazolinone	isothiazolinone
Mode of action	electrophilic active agent with activated N-S bond; reacts with nucleophils (e. g. amines or thiols); enzyme inhibition, destruction of protein thiols, generation of free radicals	electrophilic active agent with activated N-S bond and vinyl activated halogens; reacts with nucleophilic elements of cell proteins resulting in interruption of important metabolic processes
Activiy	excellent efficacy against a wide range of fungi; covers also diverse bacteria species	provides a broad antimicrobial activity against fungi, yeasts, bacteria, algae and sea animals at very low inhibitory concentrations
Approximate dosage in the coatings area [ppm]	250 – 500	100 – 1000
Remarks	sensitizer; low leaching resistance	strong sensitizer; wetstate stability in some cases limited

encing factors not only can substantially lower the concentration of algicidal (as well as fungicidal) actives in the upper layer of a cured paint film due to degradation or chemical transformation processes, but they may also ultimately lead to severe disintegration of the surface to be protected (see Figure 2.23).

Algae and cyanobacteria are often quoted in the literature as belonging to the primary colonisers on surfaces. Some authors claim that algae together with cyanobacteria are the first colonisers that prepare the ground for heterotrophic organisms, such as fungi[111–114], whereas others argue that cyanobacteria are among the majority of the first organisms that condition the surfaces for microbial coverings, such as green algae, lichens, and mosses[115, 116, 119, 120].

BBIT	**Tebuconazole**	**Propiconazole**
4299-07-4	107534-96-3	60207-90-1
brown liquid	white to beige powder	yellowish viscous liquid
no obvious odour	slight characteristic odour	odourless
~ 95.5–97 %	≥ 95 % (racemic mix. 1 : 1)	≥ 96 % (mix. 4 stereoisomers)
	105 °C (99.9 % purity)	
decomp. before boiling		> 250 °C (92.2 % purity)
1.5 x 10^{-2} (25 °C)	1.7 x 10^{-6} (20 °C)	5.6 x 10^{-5} (25 °C, 99.1 %)
760–800 (pH 7, 10–30 °C)	29 (20 °C, pH 7)	100 (pH 6.9, 20 °C, 99.1 %)
2.86 (pH 6.7-6.9, 25 °C)	3.49 (20 °C)	3.72 (pH 6.6, 25 °C, 99.1 %)
6, 7, 9, 10, 13	7, 8, 10	7, 8, 9
–	7, 8, 10	7, 8, 9
Czech Republic	Denmark	Finland
isothiazolinone	azole	azole
electrophilic active agent with activated N-S bond; affects transport processes across the cytoplasmic membrane and the electron transport chain utilized by bacteria	inhibition of ergosterol biosynthesis in fungal membranes; at lower azole conc. retardation of fungal growth, at higher azole conc. disruption of fungal membranes	inhibition of ergosterol biosynthesis in fungal membranes; at lower azole conc. retardation of fungal growth, at higher azole conc. disruption of fungal membranes
provides a broad efficacy spectrum against fungi, yeasts and a couple of algae species; covers also a range of bacteria, but here with only moderate action against *Pseudomonas spp.*	high efficacy against wood destroying fungi (Basidiomycetes); some gaps against blue-staining fungi; often applied in the wood preservation area in particular in combination with IPBC	activity spectrum similar to that one of tebuconazole (see left); often applied in the wood preservation area in particular in combination with IPBC
1000–3000 (PT 7) 250–500 (PT 6)	1500–5000	1000–3000
sensitizer; chemically and thermally stable	approved under European plant protection legislation; candidate for substitution under BPR (meets 2 of the PBT criteria)	proposed for Repr. 1B (H360D) classification (exclusion criterion under BPR), sensitizer

The range of suitable algicides to choose from is even more limited than that for fungicides for film preservation (see also Section 2.2.1). More or less four declared algicidal actives are currently available, while the corresponding dossiers are currently being assessed that will enable them to remain on the market under transitional provisions of the European biocides legislation[126, 181]. The chemistries behind these active substances (see also Table 1.3, Section 1) are typified by phenylurea derivatives, triazine compounds and electrophilic active agents[133, 158, 159, 164–169]. For the majority of these, the mode of action involves interference with the photosynthesis process (see also Table 2.7), which is the basic food source for algae. It has already been mentioned in Section 2.2.1 that the two fungicides IPBC and zinc pyrithione also possess a certain degree of algicidal activity. However, their performance here is not very pronounced and so they are typically combined with the algicidal actives described on the following pages. The chemical substance cybutryn is only mentioned here for the sake of completeness, as it was frequently employed in film preservation in the past. It is now no longer supported under BPR, however. Figure 2.24 provides an overview of algicides frequently used in coatings, some of which were originally developed for plant protection.

2.2.2.1 Diuron

The active substance diuron (IUPAC name: 3-(3,4-dichlorophenyl)-1,1-dimethylurea, CAS No. 330-54-1) (Figure 2.24) is a whitish powder with a slight odour of ammonia or amine. The purity of the technical grade is specified as 98 to 99 %. Diuron has a melting point of approx. 156 °C (98.8 % purity), an octanol/water partition coefficient (log P_{ow}) of 2.89 at 20 °C and pH 7, a vapour pressure of 7.6 x 10^{-7} Pa (20 °C, 98.8 % purity) and a solubility in water of approx. 29 ppm (20 °C, pH 7, 98.9 % purity). The substance is moderately soluble in diverse organic solvents, such as acetone, 1,2-dichloroethane, and ethyl acetate. Diuron is not considered to be highly flammable or explosive[244, 245].

In the context of European biocide legislation, applications to approve diuron for use in the following product types are being assessed: 7 (film preservatives) and 10 (construction material preservatives)[217].

Diuron is approved under European plant protection legislation[314].

According to the harmonised classification and labelling indicated in Annex VI of the CLP Regulation[196], the active substance is classified with Carc. 2 (H351: suspected of causing cancer), with Acute Tox. 4 * (H302: harmful if swallowed) and STOT RE 2 * (H373**: causes damage to organs through prolonged or repeated exposure). The active is not considered to be a skin or eye irritant and nor is it deemed a skin sensitiser. Furthermore, there is no evi-

Figure 2.23: Examples of environmental impact on coating components

dence of genotoxicity or of adverse effects on reproductive parameters under the conditions of available studies. Animal studies reported developmental toxicity and carcinogenicity.

The PBT Expert Working Group of the Technical Committee of New and Existing Chemicals has agreed that diuron should be deleted from the list of potential PBT/vPvB substances[247]. The active substance is also not expected to be bioaccumulative[244, 245].

Diuron interferes with the algal photosynthesis process by interacting with a specific-binding site on a protein subunit of photosystem II. It does not possess bactericidal efficacy and its fungicidal properties are insignificant[133, 165–167, 169]. Diuron offers a very broad spectrum of algicidal efficacy and is usually combined with one or more declared fungicides for film preservation purposes to cover all relevant microorganisms known to settle and grow on exterior coatings, such as water-borne paints (e.g. those based on polyvinyl acetate, acrylate or styrene acrylate binder) as well as mineral renders, and organic resin/silicone resin renders. The typical dosage lies in the range 500 to 1500 ppm active substance. Diuron also delivers excellent performance even in challenging regions with a tropical climate, where higher dosages are usually employed.

2.2.2.2 Isoproturon

The active substance isoproturon (IUPAC name: 3-(4-isopropylphenyl)-1,1-dimethylurea, CAS No. 34123-59-6) (Figure 2.24) is a white to slightly yellowish solid with no significant odour. The purity of the technical grade is specified as 97 to 98 %. Isoproturon has a melting point of 157 to 159 °C (> 99 % purity), an octanol/water partition coefficient (log P_{ow}) of 2.50 at 25 °C (> 98 % purity), a vapour pressure of 1 x 10^{-6} Pa (20 °C, 99 % purity) and a solubility in water of approx. 70 ppm (20 °C, 99 % purity), the last of these being the highest of all the algicides described in this section. The corresponding value (see above) is more than twice as high as that for the solubility in water of diuron and terbutryn (29 ppm and 25 ppm

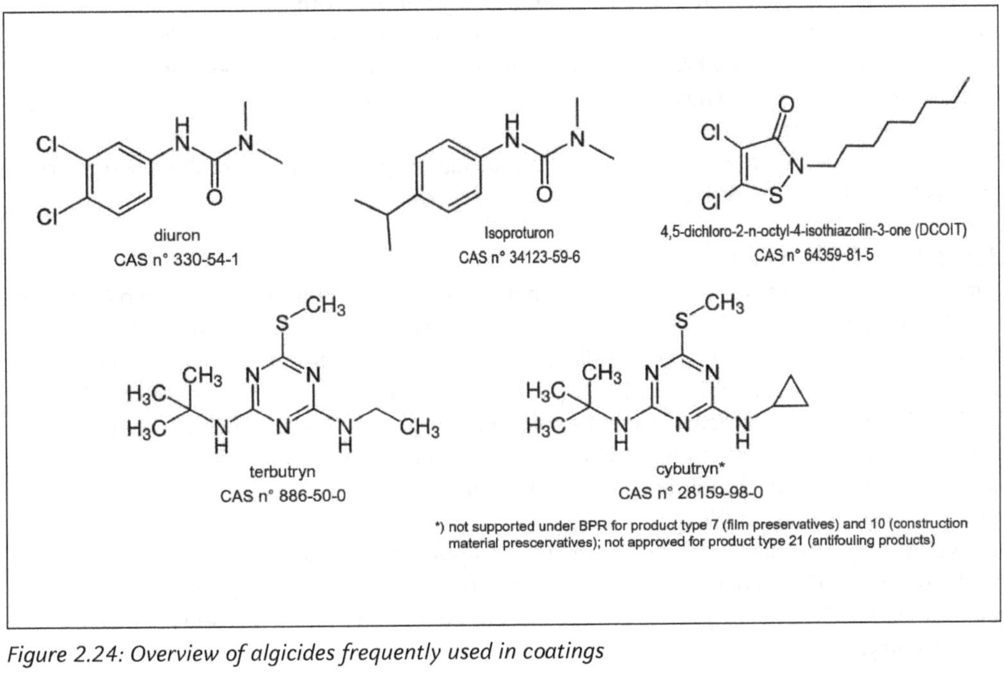

Figure 2.24: Overview of algicides frequently used in coatings

respectively). At 3.5 ppm, DCOIT has the lowest solubility in water. Isoproturon is quite readily soluble in diverse organic solvents, such as acetone, 1,2-dichloroethane, and methanol. The substance is considered to be neither highly flammable nor explosive nor oxidising[248].

In connection with the European biocide legislation, applications to approve isoproturon for use in the following product types are being assessed: 7 (film preservatives) and 10 (construction material preservatives)[217].

Approval of the active substance isoproturon under Plant Protection Products Regulation was not renewed after 1 July 2016[246]. Member States were required to have withdrawn authorisations for plant protection products containing isoproturon as active substance by 30 September 2016. Any grace period granted by Member States expire on 30 September 2017 at the latest.

The substance is not considered to be a skin or eye irritant and nor is it deemed a skin sensitiser. Furthermore, there is no evidence of genotoxicity. Within the framework of the harmonised classification and labelling process (CLH) throughout the European Union, a dossier has been submitted by the German Competent Authority[249]. The proposed future entry in Annex VI of CLP Regulation[196] provides for classification in relation to carcinogenicity (Carc. 2, H351: Suspected of causing cancer), to reproduction toxicity (Repr. 2, H361f: Suspected of damaging fertility) and with STOT RE 2 (H373: May cause damage to organs (blood) through prolonged or repeated exposure).

In the event that the proposed classification and labelling of isoproturon with Carc. 2 and Repr. 2 is confirmed by the authorities, this would mean that the so-called 'interim criteria' for endocrine-disrupting properties are met. Substances possessing such properties would fall under the exclusion criteria[126] of Article 5 (3) of the BPR: "Pending the adoption of those criteria, active substances that are classified in accordance with Regulation (EC) No. 1272/2008 as, or meet the criteria to be classified as carcinogen category 2 and toxic for reproduction category 2, shall be considered as having endocrine-disrupting properties."

Like diuron, isoproturon interferes with the algal photosynthesis process by interacting with a specific-binding site on a protein subunit of photosystem II[133, 165–167, 169].

The active possesses a broad spectrum of algicidal efficacy, but does not exhibit bactericidal or fungicidal properties[127]. It is usually combined with film fungicides, but there are also commercial formulations of isoproturon with other algicides, such as combinations with terbutryn (see below) in the ratio 10 : 1. The typical dosage for isoproturon in dry-film applications lies in the range 500 to 1500 ppm active substance.

2.2.2.3 Terbutryn

The active substance terbutryn (IUPAC name: 2-tert-butylamino-4-ethylamino-6-methylthio-1,3,5-triazine, CAS No. 886-50-0) (Figure 2.24) is a white powder with a slight odour. It has a melting point of approx. 104 °C, an octanol/water partition coefficient (log P_{ow}) of 3.74, a vapour pressure of 2.3 x 10^{-4} Pa (25 °C) and a solubility in water of approx. 25 ppm (20 °C). The substance is readily soluble in diverse organic solvents, such as acetone, n-octanol, and methanol. Terbutryn is considered to be neither flammable nor self-igniting and is stable under normal conditions. The methylthio group is hydrolyzed in the presence of strong acids or alkalis[250, 251].

With regard to European biocide legislation, applications to approve terbutryn for use in the following product types are being assessed: 7 (film preservatives), 9 (fibre, leather, rubber and polymerised materials preservatives) and 10 (construction material preservatives)[217].

Terbutryn is not approved under European plant protection legislation[314].

Currently, there is no harmonised classification and labelling of terbutryn under the CLP Regulation[196]. According to the ECHA Classification & Labelling Inventory[252], most entries indicate self-classification with Acute Tox. 4 (H302: harmful if swallowed) and with Skin Sens.1 (H317: May cause an allergic skin reaction). The latter triggers EUH208 labelling on the product to be protected with 'Contains terbutryn. May produce an allergic reaction' at concentration levels higher than 1000 ppm. Moreover, classification for aquatic toxicity also exists which entails labelling of respective product formulations at relatively low dosage levels equal to or higher than 250 ppm with the hazard symbol GHS 09 ('dead fish, dead standing tree').

Reported effects of exposure to terbutryn[250] include toxicity by all routes (inhalation, ingestion, dermal contact) as well as skin and eye irritation. The Annual Cancer Report (2015) from the U.S. Environmental Protection Agency (EPA), Office of Pesticide Programmes, classifies terbutryn as 'Possible Human Carcinogen – Group C'[253].

The mode of action of terbutryn involves inhibition of photosynthesis by disruption of light reactions and blocking of electron transport[254].

The active is highly stable, even in very alkaline environments, such as silicate paints. Terbutryn provides generally high activity against a variety of common algae, but exhibits some imbalance in its efficacy against selected species[127]. Leachability of the active is low, but the substantial vapour pressure (akin to that of DCOIT) could pose a potential risk for depletion of the active from coating due to evaporation. The typical dosage for use of terbutryn as algicide in coatings is 200 to 1000 ppm.

2.2.2.4 DCOIT

The active substance DCOIT (Figure 2.24) has already been discussed in Section 2.2.1.6. because the substance possesses also good fungicidal effectiveness in addition to its algicidal properties. Moreover, DCOIT is also employed as a broad-spectrum cobiocide in antifouling paints for controlling soft-fouling (e.g. fungi, algae, diatomaceous slimes). These broad antimicrobial properties would normally qualify DCOIT as a stand-alone film preservative in coatings. However, its strongly sensitising/strongly irritating properties represent a significant downside, since using it, even at very low dosage levels, entails classification and labelling of the product to be protected (see also Section 2.2.1.6).

2.2.2.5 Summary of important properties of film algicides

Table 2.7 summarises and contrasts important properties of the described algicidal actives frequently used in coatings for dry-film preservation purposes.

2.2.3 Overview of fungicidal/algicidal product formulations

As already mentioned in the previous sections, microbicidal product formulations intended for film preservation purposes are usually mixtures of various fungicidal and algicidal actives (see Table 2.8). The balancing act here lies on one hand in trying to cover all relevant germs and on the other in keeping the dosage levels of the microbicides as low as possible to avoid triggering classification and labelling of the products to be protected. A strategy of employing active substance combinations can also help to minimise the risk of resistance formation by the microbial species concerned, provided that the active substances employed have different modes of action.

Table 2.7: Contrasting important properties of algicides for film preservation

	Diuron	**Isoproturon**
CAS n°	330-54-1	34123-59-6
Physical state	whitely powder	white to slightly yellowish solid
Odour	slight ammonia/amine odour	odourless
Technical grade a.i.	~ 98–99 %	~ 97–98 %
Melting point	~ 156 °C (98.8 % purity)	~ 157–159 °C (> 99 % purity)
Vapor pressure [Pa]	7.6 x 10^{-7} (20 °C, 98.8 % purity)	1 x 10^{-6} (20 °C, > 99 % purity)
Water solubility [ppm]	29 (20 °C, pH 7, 98.9 %)	70 (20 °C, > 99 % purity)
log P_{OW}	2.89 (pH 7, 20 °C)	2.50 (25 °C, > 98 % purity)
Application for approval (PT) under EU biocide legislation	7, 10	7, 10
Approval (PT) under EU biocide legislation (as per May 2017)	–	EU wide phase out on 30.09.2017 at the latest in the field of crop protection
Evaluating Competent Authority	Denmark	Germany
Approval under EU plant protection legislation (Expiry Date)	yes (30.09.2018)	yes (30.06.2016)
Chemistry	phenylurea derivative	phenylurea derivative
Mode of action	mode of action is related to a specific-binding site on a protein subunit of photosystem II thereby impacting the photosynthesis process	mode of action is related to a specific-binding site on a protein subunit of photosystem II thereby impacting the photosynthesis process
Activity	highly effective against both sea-water and freshwater algae; bactericidal and fungicidal properties are insignificant	provides a broad spectrum of algicidal efficacy; does not provide bactericidal or fungicidal properties
Approximate dosage in the coatings area [ppm]	500–1500	500–1500
Remarks	not a PBT/vPvB substance; classified with Carc. 2, light stable	proposed future classification triggers BPR cut-off criteria for endocrine disrupting substances

The number of possible combinations is fairly high and the choice of a specific microbicidal product formulation depends on different factors, such as performance in laboratory/semi-field pre-trials (see Section 4), rejection of certain chemistries due to human health/environmental issues (e.g. sensitising effects, CMR substances), possible interactions of one or more active substances with coating ingredients (e.g. discolouring problems) and, finally, preservation costs.

Terbutryn	DCOIT
886-50-0	64359-81-5
white powder	off-white solid
slight odour	odourless
not available	~ 98–99 %
~ 104 °C	~ 41–42 °C
2.3 x 10^{-4} (25 °C)	9.8 x 10^{-4} (25 °C)
25 (20 °C)	3.5 (pH 7, 20 °C)
3.74	2.80 (pH 7, 23 °C)
7, 9, 10	7, 8, 9, 10, 11, 21
–	8, 21
Slovakia	Norway
no	no
triazine derivative	isothiazolinone
inhibition of photosynthesis by disrupting light reactions and blocking the electron transport	electrophilic active agent with activated N-S bond and vinyl activated halogens; reacts with nucleophilic elements of cell proteins resulting in interruption of important metabolic processes
provides in general high activity against a variety of common algae; some imbalance in activity against selected species	provides a broad anti-microbial activity against fungi, yeasts, bacteria, algae and sea animals at very low inhibitory concentrations
200–1000	100–1000
sensitizer (EUH208 at conc. levels >1000 ppm); indicated as possible human carcinogen (group C) by US EPA	strong sensitizer; suggested classification as skin sensitizer with a specific concentration limit (SCL) of 10 ppm

2.3 Plant hygiene

Industrial facilities generally and coatings production plants in particular mostly consist of diverse components that are joined together in a complex manner. These include reactor vessels, piping systems, tanks, product containers, dosage units, pumps, valves, filters, filling devices and the like (for examples, see Figure 2.25).

Table 2.8: Frequently employed film preservation combinations – generic illustration (fig. in the chart represent typical active contents (or their ranges) in wt. %)

	Active ingredients	BCM	IPBC	OIT	TBZ	ZnP	Propiconazole	DCOIT	Diuron	Isoproturon	Terbutryn
Fungicide Combos	BCM + OIT	3–30		2–10							
	BCM + IPBC	15	5								
	IPBC + OIT		30	20							
	IPBC + propiconazole		22–23				7–8				
	TBZ + OIT			4	25						
	ZnP + BCM	5–7				5–10					
	ZnP + OIT			2–7		2–7					
	ZnP + OIT + IPBC		3	3–5		5					
Fungicide/Algicide Combos	BCM + diuron	5–10							10–20		
	BCM + terbutryn	10									10
	IPBC + DCOIT		6–13					13–27			
	IPBC + diuron		10						5.5		
	IPBC + terbutryn		7–14								8–14
	IPBC + terbutryn + isoproturon		7							12	1.2
	OIT + DCOIT			3				10			
	OIT + diuron			10					10		
	OIT + terbutryn			6–10							10
	ZnP + terbutryn					6–20					2–10
2 Fungicides/Algicide Combos	BCM + IPBC + diuron	9	3						15		
	BCM + IPBC + terbutryn	9–12	3–4								4–10
	BCM + IPBC + terbutryn + isoproturon	9	3							8.5	0.85
	BCM + ZnP + terbutryn	6				10					5 – 6
	OIT + BCM + diuron	5–10		2–5					10–22		
	OIT + BCM + terbutryn	5–10		2–4							5–10
	OIT + DCOIT + terbutryn			6–7				1–2			17–18
	OIT + IPBC + diuron		4–5	3–4					14–15		
	OIT + IPBC + terbutryn		9	3							5
	OIT + propiconazole + terbutryn			2–3			15				10
	OIT + TBZ + terbutryn			4	2–3						5
	OIT + ZnP + diuron			2–5		4–8			10–15		
	OIT + ZnP + terbutryn			1–8		4–21					2–15

Not surprisingly, sophisticated industrial complexes can be susceptible to microbial attack when they are used to process water-borne products. Microorganisms, such as bacteria, fungi and yeasts, are remarkably adaptive to different environmental situations and they can also find their specific ecological niche here (see Figure 2.25 for examples). Even materials which have long been thought to withstand microbial degradation, such as plastics and concrete, are not exempt from such attack[1]. This is where the formation of biofilms plays a decisive role. These often occur on boundary layers, e.g. on the water surface or the interface with a solid phase (see also Section 1.1.2). Clearly, such conditions also frequently occur in industrial facilities.

Microorganisms, their segregation products (e.g. polysaccharides, proteins, lipids, nucleic acids) and water as the main component form slimy layers that have a more or less stable shape. As different microbial strains can occur in a biofilm, it is possible for aerobic germs to coexist with anaerobic germs within a distance of just a few hundred micrometres of each other. Channels, pores and cavities in the biofilm matrix allow for mass transfer and water supply[59, 60] and even intercellular communication ('quorum sensing') is known to take place between certain species[61, 62] by means of which specific signalling molecules support the coordination of a set of processes in biofilms.

A biofilm provides effective shelter against environmental stress, such as substantial pH or temperature fluctuations and even confers a certain degree of protection against microbicides. The biofilm might withstand penetration by chemical molecules, thus impairing the desired antimicrobial action. Stagnating or insufficient water flow supports the formation of such films from which germs can be constantly released and flushed into the piping system where they may initiate further contamination in other parts of the production facility. Consequently, optimum flow characteristics in piping systems are essential if germ proliferation is to be avoided. Ideally, this aspect will be addressed while construction of a new plant is at the planning stage.

Figure 2.25: Examples of production plant equipment

In particular, pathogenic microorganisms, such as *Legionella* species are detected quite frequently in biofilms, where they are in a position to develop and to proliferate[255]. Their occurrence in plant facilities can severely disturb the production process, even to the extent of shutting down the factory[256].

A further consequence of biofilm formation and its unhindered proliferation is the generation of aggressive chemical compounds (such as acids) via microbial degradation processes that may ultimately lead to destruction of the colonised substrates (biocorrosion). Even alloy steel is not exempt from such attack. As a consequence of this microbially-induced corrosion, the service life of a production site can be significantly shortened. When potential shutdown times and higher energy consumption are considered as well, this phenomenon is estimated to be responsible for financial losses in the double-digit billion euros range[65].

Biofouling is a preliminary stage of bio-corrosion. Macroscopic effects on substrates can take the form of discolouration, clogging, slime formation, gas evolution and odour release and are due to biofilm proliferation. Although this biofouling in theory impairs performance, it does not yet lead to material damage[65].

2.3.1 Prevention is better than cure

Effective plant hygiene is essential for ensuring a smooth production flow in the long term. What is needed is a comprehensive programme comprising a set of measures rather than use of microbicides exclusively which, although an important part of treatment, should not constitute the only treatment. Preservatives are designed to protect clean products against occasional contamination, not to withstand repeated contact with high levels of germs. Microbicides can only provide the necessary protection when proper plant hygiene measures are correctly implemented in parallel. Critical sources of contamination need to be identified and possible transmission paths of microbiological infestation in the manufacturing process need to be

Figure 2.26: Examples of microbial contamination of production plant components

interrupted. Systematic hygiene measures must be accompanied by constant monitoring to ensure that microbial contamination is detected immediately and that remedial action can be initiated as necessary and as soon as possible. This safeguards the manufactured products and protects the employees against pathogenic germs and so ultimately also creates a socioeconomic benefit.

Many preservatives are consumed in the process of combating microorganisms and their concentration can be depleted to sublethal levels. Underdosing of preservatives can lead to the development of adapted microbiological populations (microbial resistance) and should therefore be strictly avoided.

There are many conceivable points of attack in production (see also Table 2.9). Not only is cleanliness in the workplace essential (this includes all operational equipment, e.g. reactor vessels, piping systems, tanks, product containers, dosage units, pumps, valves, filters, filling devices), but susceptible raw materials also must be checked on a regular basis (e.g. polymer emulsions, pigment slurries) and personnel need regular training in this area.

Table 2.9: Examples of plant components susceptible to microbial attack

Location	**bacteria count [CFU/g]**
Fast-rotating agitator in paint emulsion	1×10^5
Storage tank, emulsion	4×10^3
Kneader	4×10^4
Strainer, emulsion	2×10^5
Funnel tube	1×10^4

CFU: colony-forming unit; unit to estimate the number of viable germ cells in a sample, Source: H. Brill, Mikrobielle Materialzerstörung und Materialschutz, Gustav Fischer Verlag Jena 1, pp. 1–290 (1995)

Table 2.10: Examples of raw materials that serve as a potential nutrient base for germs

Paints	**Adhesives**	**Lubricoolants**
thickeners defoamers dispersants pigments polymer emulsions (binders) ammonia	starch corn starch (modified) other thickeners stearic acid urea defoamers polymer emulsions	mineral oil emulsifiers defoamers corrosion inhibitors EP additives dyestuffs fragrances
pH 8–9 (approx.)	pH 6.5–7.5 (approx.)	pH 9–10 (approx.)

2.3.2 Where there is water, there is also life

Employees should commit to taking responsibility for their own workplaces and to ensuring that the plant components in those areas are kept clean and dry. Ideally, production equipment will be designed for easy access. Spilled materials and waste must be removed immediately. Container openings, hatches and the like should be kept closed for as long as possible because ambient air contains all kinds of bacteria and fungal spores. The latter are extremely light and can be easily transported on draughts of air to other areas in the facility, where they can contaminate locations that were thought to be clean and free of germs. Raw materials and finished products can potentially serve as a nutrient base for microorganisms (see Table 2.10), enabling rapid proliferation, especially when the ambient temperature and humidity fits the requirements of the germs in question. Therefore, unnecessary air currents should be avoided and appropriate air filters should be installed. Filter cartridges must be checked regularly and replaced as necessary.

Table 2.11: Possible sources of microbial contamination in coatings production

Air	**Water**	**Raw materials (solid, powdery)**	**Raw materials (liquid, aqueous)**	**Plant design & other factors**
air conditioner air filter air draft	tap water fountain water river water storage tank water wash water	fillers pigments thickeners	polymer emulsions pigment dispersions additives	biofilm formation local infestation "dead points" contamination via personnel

Microbial limits should be specified for susceptible raw materials and verified prior to use. Dust spreading from powdered raw materials must be minimised. The need for possible measures to preserve aqueous, liquid raw materials should be discussed with suppliers and contaminated products exhibiting signs of microbial infestation must be tested for serviceability (e.g. after curative treatment with microbicides) and, where necessary, discarded. Ideally, these measures will form an integral part of a quality management system.

There are numerous ways for germs to infiltrate production facilities (see Table 2.11). These include not only the raw materials but also the air and water supplies, the installation of plant components and contamination by personnel.

With regard to plant design, the materials employed should be as resistant as possible to microbial attack and should also be easy to clean (e.g. stainless steel installations and equipment can be steam-cleaned). The number, lengths and bends of pipes should be kept to a minimum, as well as the number of joints, valves, gauges and the like. 'Dead-ends' should be avoided or eliminated (see Figure 2.27).

Ideally, the pipes will be installed under a gradient for better drainage and to minimise stagnation of water within the distribution system. Cleaning of piping systems (particularly

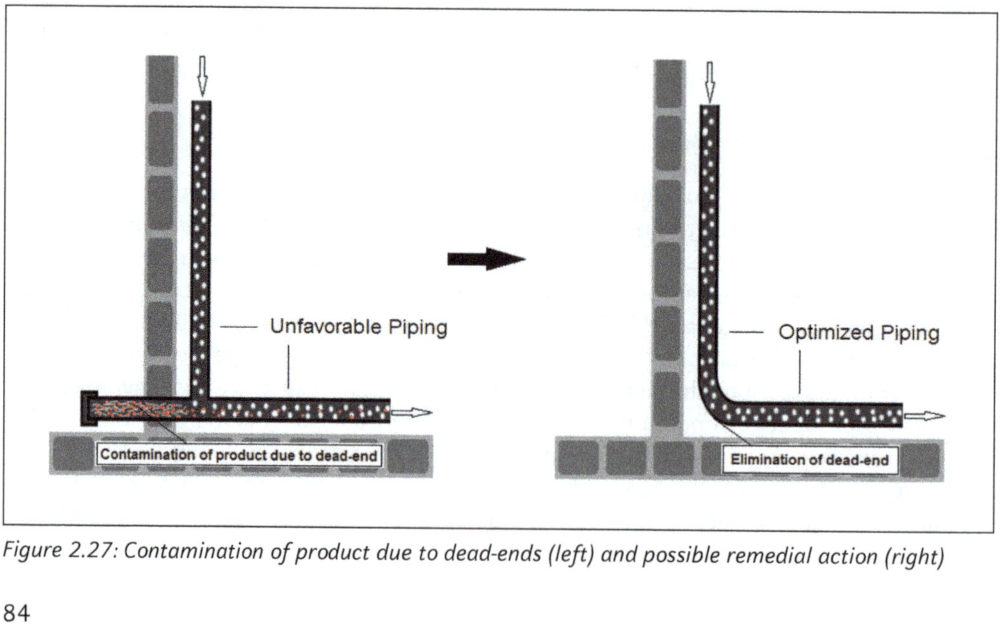

Figure 2.27: Contamination of product due to dead-ends (left) and possible remedial action (right)

those used for product filling) should be performed regularly. Sagging loops can lead to build-up of dangerous sources of contamination and should therefore be avoided (see Figure 2.28).

The best storage tanks have a conical shape for effective flow, even when there is little in the tanks, and for easier removal of residuals. However, tank headspace should be minimised whenever possible. Ideally, tanks will be filled from the top with fresh raw material and discharged at the bottom (first in, first out). Additionally, they will be equipped with an effective agitator and a wetting system to prevent liquids from forming a skin on the surface. Spy-glasses and sampling points facilitate monitoring while proper thermal insulation will protect the tank contents from elevated temperatures that would otherwise favour the proliferation of germs.

Regular cleaning and disinfection of the tanks and corresponding equipment are indispensable. All aggregations, incrustations, sedimentations or attachments should be removed (e.g. by scraping, power steam-cleaning or alkali soak) down to the bare surface, as any residues can cause repeated re-infection of each new production batch. Small spots can be treated with hydrogen peroxide solution. These recommendations also apply to reactor vessels, as well as to piping and other equipment, wherever this is possible and there is access. Cleaning water should not be returned to the production process. Subsequent drying of unused plant components avoids unnecessary condensation that could be the starting point for re-infestation with microorganisms.

Process water and any stored or recycled water must be monitored for its microbiological quality (check for germ count) and, if necessary, treated with an appropriate microbicide or, for instance, exposed to a UV radiation unit. Stagnant water in the distribution system should be kept to a minimum or, best of all, completely avoided (flow-circulation measures might be considered here). Scheduled microbicidal treatment should be performed as early as possible in the production process. Disinfection of water-treatment units, such as ion exchange resin beds, reverse osmosis systems and filters, must be performed regularly.

In the field of plant hygiene, the actives used for disinfection/preservation can be categorised as oxidisers and non-oxidisers[257]. Oxidising substances offer *inter alia* broad-spec-

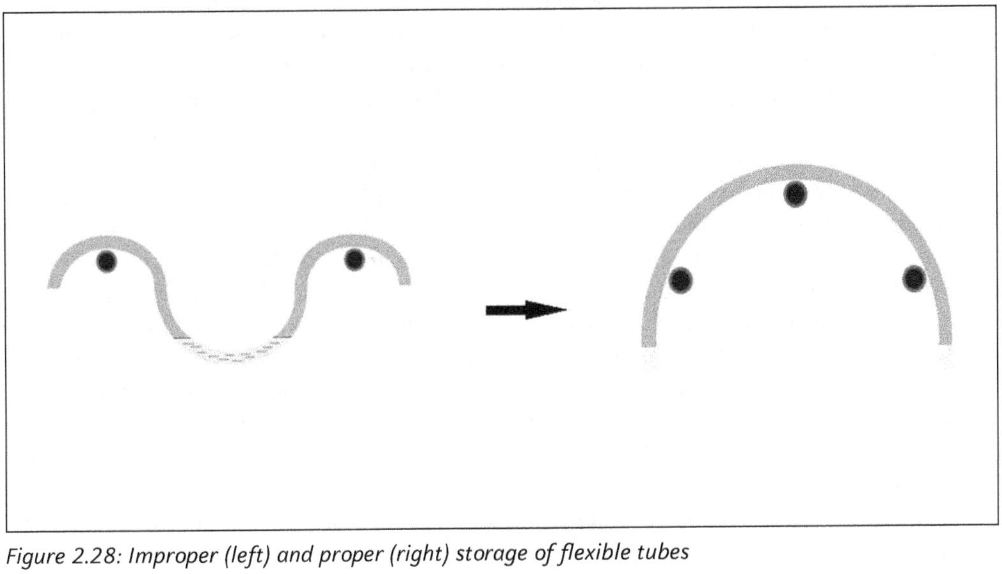

Figure 2.28: Improper (left) and proper (right) storage of flexible tubes

Table 2.12: Example of microbicide dosage in closed-loop cooling

Oxidizers

Active	Dosage active ingredient [ppm]
(Active) chlorine	0.1–0.2
Sodium hypochlorite	1–5
Ozone	0.015–0.2
Hydrogen peroxide	1–50

Non-Oxidizers

Active	Dosage active ingredient [ppm]
Glutaraldehyde	1–200
DBNPA	1–50
BNPD	1–25
Isothiazolinones	1–30
Quaternary ammonium compound	3–50

trum efficacy against bacteria, fungi and algae, along with rapidity and various modes of action against germs[139, 258]. As this class of microbicides acts at very low ppm levels in industrial waters, they make a very attractive, cost-effective preservation proposition for industry. A further advantage is their efficacy against biofilms and critical germs, such as *Legionella spp.*[259–262]. However, the high reactivity of oxidisers can also harbour some disadvantages, such as susceptibility to process contaminants, corrosion issues, and the risk of accidental reaction with other chemicals in production[257].

The properties of non-oxidising microbicides, such as BNPD, DBNPA, isothiazolinone derivatives and formaldehyde-releasing compounds are discussed in Section 2.1. Typical dosage levels in the field of plant hygiene are indicated in Table 2.12 for an example taken from the field of closed-loop cooling[263].

2.3.3 Ten-point programme: disinfect operational facilities

The following ten-point programme is recommended (this may need to be adapted to the particular case) as an example of how to disinfect operational facilities:

1. Wherever accessible, all visible aggregations, incrustations, sedimentations or attachments are removed down to the bare surface (e.g. by scraping, power steam-cleaning or alkali soak). Small spots are treated with hydrogen peroxide solution.
2. Vessels, tanks, piping systems, installations and the like are flushed with either hydrogen peroxide or sodium hypochlorite solution. (Both of the above are highly cost-effective disinfection formulations that also cover biofilms and critical germs, such as *Legionella spp.)*
3. In places where the use of such oxidisers under 2) might be critical (corrosion!), non-oxidising microbicidal formulations, such as DBNPA, BNPD, isothiazolinones, glutardialdehyde (GDA) or quaternary ammonium salt solutions are possible alternatives.
4. Flow-circulation measures during cleaning are ideal, where applicable and feasible.
5. If possible, microbicidal treatment should allow for several hours' dwell time.
6. All treated plant installations must be thoroughly flushed with adequate amounts of **clean** water. Subsequent drying is not imperative for those plant components which are required immediately for new production activities; otherwise, drying is recommended to minimise aqueous residuals which might serve as a point of attack by future contaminants.
7. Production waters used for dilution will ideally contain an appropriate in-can microbicide (see Section 2.1).

8. Cleaning and disinfection measures should be carried out regularly basis ("A stitch in time saves nine").
9. Appropriate protective clothing must be worn by personnel during disinfection measures.
10. All industrial safety legislation and specifications must be met.

Also recommended are regular plant-hygiene audits as a way of monitoring the hygiene situation in a manufacturing site. For this, the following action points are worth mentioning:
- Proper preparation for the plant audit
- Identification of already known contamination issues and of probable contamination spots
- Sampling of industrial fluids, process waters, raw materials and finished products
- Taking of swabs from spots where process intermediates or final products are processed (e.g. tank inlets, filling devices)
- Application of aseptic techniques by experienced staff
- Repeated sampling at the same spots to provide comparative data
- Proper documentation of findings
- Recommendations for hygiene measures (remedial action, curative treatment)
- Regular follow-up inspections

Biocidal product suppliers usually offer a plant hygiene service package which includes the audit (visual inspection, sample taking, training of production personnel etc.), a determination of the sample germ count (sterility check) in the supplier's lab, identification of the microorganisms involved, where applicable and where desired, and finally documentation containing all the germ count test results, conclusions as well as recommendations (typically provided as a report).

3 Application aspects

Microbicidal product formulations are widely used in various fields of industrial preservation. The most obvious applications in the coatings area are the protection of water-borne products in the wet state (e.g. paints, pigment slurries, binder emulsions) against bacteria, fungi and yeasts (PT 6 under BPR (Biocidal Products Regulation)) and the preservation of paint films against fungal and algal infestation after the liquid product has been applied and cured on the substrate to be protected (PT 7 under BPR). But coatings are not limited to such products. Plasters, renders, mortars, sealants, plastics and the like also form surfaces which can be susceptible to microbial attack. Annex V to the BPR[126] provides *inter alia* the following definitions:

PT 7 (film preservatives)
- "Products used for the preservation of films or coatings by the control of microbial deterioration or algal growth in order to protect the initial properties of the surface of materials or objects, such as paints, plastics, sealants, wall adhesives, binders, papers, art works."

PT 10 (construction material preservatives)
- "Products used for the preservation of masonry, composite materials, or other construction materials other than wood by the control of microbiological, and algal attack."

These BPR definitions offer some scope for interpretation. There have already been discussions in the past as to whether, for instance, preservative formulations designed for protecting plasters or renders fall under PT 7, as such substrates can be interpreted as thick-layer coating films, or whether they come under PT 10. The background to this discussion was that assessment by the authorities for use in two product types could conceivably incur more costs (e.g. for fees, dossier preparation, efficacy data) than for use in just one. However, to avoid disagreeable surprises, applications have been submitted for approval to use most of the actives intended for film preservation in both PT 7 and PT 10 (see also Table 2.5, Table 2.6 and Table 2.7).

A further borderline case is the protection of wood substrates. Products which 'only' create a superficial surface layer on such matrices and which do not claim to provide deep preservation within the substrate are regarded as wood coatings and as such do not fall under the definition of a biocidal product under BPR. However, product formulations containing microbicidal actives which penetrate wood structures to a certain extent to protect them against wood-destroying or wood-disfiguring organisms (including insects) and which are claimed as such are seen as biocidal products that fall under PT 8 (wood preservatives).

PT 8 (wood preservatives)
- "Products used for the preservation of wood, from and including the saw-mill stage, or wood products by the control of wood-destroying or wood-disfiguring organisms, including insects."

In this context, it should be noted that the PT 8 formulations mentioned (e.g. wood primers or wood stains) will require – because they are biocidal products – their own authorisation under BPR in due time if they are to remain on the market, whereas wood coatings in the sense of PT 7 do not need such authorisation. This latter situation is quite similar to that for architecture paints containing dry-film preservatives. In this case, a formulation containing fungicides/algicides which is intended to be added to coating formulations for preservation purposes is in fact the microbicidal product and so a corresponding dossier has to be submitted for the intended use at the product evaluation stage. Consequently, the biocidal product supplier is responsible for registering his fungicide/algicide formulation under BPR. The latter is added to coating formulations during the production process. The final end-use product (e.g. a paint) is called a 'treated article' (see Section 5).

Other application areas of relevance in the field of coatings pertain to the remedial treatment of construction materials (PT 2), to the preservation of fibrous or polymerised substrates (PT 9) and to antifouling products (PT 21). The last of these can be seen as a specialised coating because it is in permanent water contact after application. The following associated definitions in Annex V of the BPR[126] apply:

PT 2 (disinfectants and algicides not intended for direct application to humans or animals)
- "Products used for the disinfection of surfaces, materials, equipment and furniture which are not used for direct contact with food or feeding stuffs.
- Usage areas include, inter alia, swimming pools, aquariums, bathing and other waters; air conditioning systems; and walls and floors in private, public, and industrial areas and in other areas for professional activities.
- Products used for disinfection of air, water not used for human or animal consumption, chemical toilets, waste water, hospital waste and soil.
- Products used as algaecides for treatment of swimming pools, aquariums and other waters and for remedial treatment of construction materials.
- Products used to be incorporated in textiles, tissues, masks, paints and other articles or materials with the purpose of producing treated articles with disinfecting properties."

PT 9 (fibre, leather, rubber and polymerised materials preservatives)
- "Products used for the preservation of fibrous or polymerised materials, such as leather, rubber or paper or textile products by the control of microbiological deterioration.
- This product-type includes biocidal products which antagonise the settlement of micro-organisms on the surface of materials and therefore hamper or prevent the development of odour and/or offer other kinds of benefits."

PT 21 (antifouling products)
- "Products used to control the growth and settlement of fouling organisms (microbes and higher forms of plant or animal species) on vessels, aquaculture equipment or other structures used in water."

Classification for use in a certain product type is a matter of relative importance, since it may impinge on the human health effects assessment, on the environmental effects assessment or on the evaluation of the effectiveness at combating target organisms according to the specifications given in the BPR.

3.1 Service life of microbicides

As a matter of course, the in-can protection afforded to paints and coating products should not only be ensured during production, but also extend far beyond during the storage period (e.g. in a paint shop), and it might be several months or even more than 1 year before the product reaches the end user. What is more, the entire contents of a paint container might not be consumed during first use, so that the rest of the product might be stored again for significant time periods, especially in the DIY area (as we probably all know from experience). The consumer usually expects the product to remain in good condition for the duration of that period. Consequently, it should withstand microbial attack in the long term, even though the paint container might have been opened several times and at the risk that germs from the environment may have discovered the product as a most welcome nutrient source.

It is very difficult to predict from theory how long a selected microbicidal formulation for in-can protection purposes can effectively keep a water-borne product in the wet state free from microbial infestation because bacteria, fungi and yeast prefer different environmental conditions, such as pH and temperature. Germs in their optimum environment might be capable of withstanding microbicidal action better than microorganisms faced with adverse conditions. From the opposite perspective, silicate paints containing highly alkaline media are less susceptible to microbial attack than are paint formulations with just a slightly alkaline pH. In terms of microbial sensitivity, it might be less challenging to store a water-borne coating formulation in Norway than in the south of Spain, due to the different temperature conditions, especially in summer time.

Experience shows that combinations of isothiazolinones (especially CMIT/MIT + BIT) and formaldehyde releasers provide excellent efficacy, even over the relatively long term (see also Table 2.4). However, the trend for preferring microbicidal products that possibly have less impact on human health (no sensitiser, formaldehyde-free etc.) is making it more and more difficult to choose between commercial products. Figure 3.1 provides a rough overview of germs' requirements for proliferation.

What is more, some coating components can incapacitate microbicidal actives, e.g. by simple adsorption, especially when their solubility in water is low, or by encasing the active substances, for instance in micelles (e.g. generated by wetting agents or surfactants). In the worst case, the integrity of a microbicidal active will be severely impacted, for example the trans-chelation processes described for zinc pyrithione in Section 2.2.1.4, or by degradation/decomposition as a result of interaction with paint constituents. It is therefore a matter of urgency that preliminary lab trials be conducted on the compatibility of the microbicidal formulation with the coating system to be protected (see Section 4).

In the field of dry-film preservation, architects and building companies are legally obliged to provide and to erect for their clients buildings which are free of defects; this especially includes coatings free of microbial infestation, both indoors and out-

	Bacteria	Fungi	Yeasts	Algae
Light	no	no	no	yes
pH	7–10	3–8	6–8	6–8
Temperature	25–40 °C	20–35 °C	20–35 °C	15–30 °C
Nutrients	C, H, N	C, H, N	C, H, N	CO_2
Oxygen	O_2, SO_4^{2-}, NO_3^-	O_2	O_2	O_2
Water	yes	yes	yes	yes

Figure 3.1: Overview of germs' requirements for proliferation

doors. In Germany, the time-frame for guarantee claims for such coating products is usually 5 years. A minimum requirement is therefore that the applied coating must withstand fungal and algal infestation during that period. It goes without saying that customers usually expect much longer service lives, e.g. 10 years and longer, because they wish to avoid the high costs of redecoration. The durability of a façade coating can be substantially influenced by the specific exposure conditions. South-facing surfaces with unhindered incident sunlight are less susceptible to microbial attack than similar surfaces which are fronted with substantial planting (shade, atmosphere enriched with organic matter; see Figure 3.2) and/or face north (Figure 3.3).

As long as film preservatives retain their efficacy against the microorganisms concerned, they will help to control algal and fungal infestation of façades. Their effectiveness will gradually decrease as they are consumed in the course of providing microbicidal action, as they undergo natural decomposition or evaporation on site, and as a consequence of leaching processes induced by precipitation. Film preservatives should therefore ideally possess low solubility in water in order that the risk of potential leaching out from the cured paint film may be minimised (see Figure 3.4).

On the other hand, the volatility of film preservatives should be as low as possible so that evaporation of the microbicidal active into the environment may be avoided. Moreover, the substances in question should be resistant to UV radiation or to oxidising and reducing agents. Several physico-chemical properties of the most prominent dry-film preservatives (fungicides as well as algicides) are discussed in Section 2.2. and summarised in Tables 2.5, 2.6 and 2.7. It is wishful thinking to expect that there exists one ideal film preservative that meets all the desired physico-chemical requirements. Diuron, for instance, is quite stable,

Figure 3.2: Microbial infestation of façades fronted by planting; left: north-west facing; right: east-facing

Figure 3.3: Microbial infestation of façades favoured by north-facing aspect

even to UV radiation, whereas zinc pyrithione is reported to be susceptible to light exposure. In contrast, it is one of the actives that has the lowest solubility in water, together with DCOIT. The stability of the latter in the wet state is quite limited, however. OIT, along with BBIT, possesses the highest solubility in water.

As there are pluses and negatives for all actives, it is only logical to combine two, three or more different film preservatives to create a well-balanced product for a specific purpose. Prominent examples of such combinations are shown in Section 2.2.3 (Table 2.8).

Blocken and *Carmeliet*[264] showed that driving rain is not usually distributed evenly on a façade, as raindrops tend to land more on the top corners and edges of a building as a result of the so-called 'wind blocking effect', while little or none falls on the central region (Figure 3.5). Consequently, only some of the total precipitation reaches the façade surface, a fact which leads to less runoff and consequently to less leaching than is theoretically possible.

The assertion that wind blocking has an important impact on leaching behaviour in practice is supported by semi-field experiments[265, 266]. The measured runoffs significantly depend on the sample size; i.e. the larger the specimens under test (representing the wall), the less experimental runoff there is relatively per square metre and, logically, the less leaching occurs (Table 3.1). It was also found that the runoff values were the most important factors influencing the leaching results in these experiments and that for the smallest test specimens (0.3 m x 0.6 m) the leaching behaviour is considerably overestimated, since the measured runoff in this special case is approximately 50 times the value for a real building 10.5 m high (see Table 3.1).

With the aid of a model house it was shown that the full quantity of rainwater does not wet the surface of a standard weathering specimen. To an extent depending on wind direc-

Figure 3.4: Two examples of leaching effects observed on exterior coatings

Figure 3.5: Two examples of the 'wind blocking effect' on façades

Table 3.1: Specimen size vs. experimental runoff – runoff from a façade decreases with increase in height due to the wind-blocking effect

Location	**Fraunhofer IME, Schmallenberg**	**Fraunhofer IME, Schmallenberg**	**EMPA, Dübendorf**	**Real Building Zürich [2] (10.5 m height)**
Test Period [Weeks]	48	48	52	52
Orientation	West	West	West	
Specimen size [m x m]	0.3 x 0.6	2.5 x 1.5	4 x (1.75 x 0.75) [1]	-
(Virtual) Wall size [m²]	**0.18**	**3.75**	**5.25**	**~ 83**
Cum. precipitation [l/m²]	901	901	815	↓
Experimental runoff [l/m²]	332	208	61	
Ratio runoff [%]	**37**	**23**	**8**	**0.7** [2]

[1] Four panels in juxtaposition with each other
[2] Data from Dr. M. Burkhardt, HSR Hochschule für Technik, Rapperswil (Switzerland)

tion and the orientation of the test specimens, on average only 40 to 50 % of the total precipitation can be collected from them. Attaching a roof overhang to the model house leads to a further substantial decrease in the amount of runoff water collected[265, 266].

Additional calculations to allow for the wind-blocking effect described above show good correlation between calculated and experimental runoff data (Table 3.2). The tests will be extended to other sites in other climatic conditions with a view to confirming these findings and to creating a theoretical basis for forecasting runoff and, what is more, predicting the leaching behaviour of a coating[265, 266].

Some further studies of the leaching behaviour of coating components have been made in recent years or are still underway[267–277]. The German Federal Environmental Agency[272] recently published a comprehensive report dealing with emissions from film preservatives into the environment. Consistent with the semi-field experiments described earlier[265, 266], it confirmed *inter alia* that:

- runoff is the most important factor influencing leaching
- cumulative emissions determined in the lab are substantially higher than those outdoors

The key points and methods of this project will now be discussed briefly (readers interested in more details should refer to the report):

A series of samples representing product types 7 (film preservatives), 9 (fibre, leather, rubber and polymerised materials preservatives) and 10 (construction material preservatives) were subjected to comparative leaching tests, both in the lab and outdoors. The project objective was to achieve a better understanding of the complex processes and to develop a method for testing the leaching behaviour of materials that might potentially be used in the context of the registration process under BPR for environmental risk assessments.

The laboratory method is based on European Standard EN 16105:2011 ("Paints and varnishes – Laboratory method for determination of release of substances from coatings in inter-

Table 3.2: Comparison of experimental and calculated potential runoff for test specimens in Schmallenberg (Germany, 2010); total precipitation: 545 l/m²; time period: 29 weeks

Panel size 2.5 m x 1.5 m	**North**	**East**	**South**	**West**	**Average**
Experimental runoff [l/m²]	69	9	31	107	54
Calculated runoff [l/m²]	71	11	37	115	59

mittent contact with water"). In such tests, coated specimens are periodically immersed in water under specified conditions. The samples are allowed to dry between dipping cycles in order to facilitate transport processes within the test materials. The leaching waters for each day of immersion are analysed to establish the identity and quantity of leached compounds[272].

Outdoor leaching tests were performed on the basis of a procedure known as NT build 509 (2005), which was originally developed for testing the leaching behaviour of treated wood.

The test specimens were oriented such that they were exposed to the maximum expected driving rain. Active substances in the comparative test were (alphabetically) carbendazim, DCOIT, dichlofluanid; diuron, IPBC, OIT, terbutryn, tolyfluanid and zinc pyrithione. Various treated articles from diverse application areas and with different chemical matrices were involved. Appropriate methods for creating suitable specimen substrates were developed. Additionally, one product was tested according to CEN/TS 16637-2:2014 ("Construction products – Assessment of release of dangerous substances – Part 2: Horizontal dynamic surface leaching test"), a method that is based on permanent contact with water. The tests were conducted at 3 different sites and the majority of test specimens were in vertical alignment and facing south-west. In a departure from NT build 509 (2005), runoff samples were taken and analysed after each rain event in the first months of the test period. Later, runoff samples were collected and then unified after a certain time period in order to streamline the analytical effort. But even then, the sampling frequency was higher than scheduled in NT build 509. Further experiments were initiated after 3 and 6 months to compare the leaching processes under different climate conditions. Finally, selected test specimens from the lab test as well as from the outdoor test were extracted and analysed for the remaining active substances in the respective matrix. Corresponding mass balances were then calculated[272].

The main conclusions in a nutshell are[272]:

- The surface areas of test specimens in outdoor testing are relatively small in relation to the façade surfaces of real buildings and the orientation of the samples under test is selected for substantial exposure to the driving rain, thus making relatively high amounts of water available for leaching and consequently creating relatively high runoff values per unit area. This means that area-related data derived from outdoor trials are not readily transferable to real building surfaces without an appropriate correction factor.
- Reproducibility of outdoor tests cannot be achieved, as leaching processes under outdoor climate conditions are influenced by a complex interrelation of different factors.
- The delineation of emission per unit area versus runoff per unit area proved suitable for comparing results from different experiments. A certain degree of variation in results must be accepted. The informational value of both test procedures (lab, outdoors) is limited with regard to the terms of use.

- Leaching from treated articles is mainly diffusion-controlled, both in the lab and outdoors (in some cases, the leaching behaviour of the active substance under test was controlled by its solubility).
- Mass balances were calculated for active substances in leachates and runoff waters, together with the remaining amounts of actives in the corresponding matrices, and the findings were correlated with the initial quantities at the outset of the test. The results revealed that, under laboratory conditions, and even more so under outdoor conditions, competing processes take place that are responsible for loss of actives, in addition to leaching. In particular, degradation/decomposition (e.g. via photolysis) and evaporation of the substances in question are believed to play a major role. Such processes can significantly outnumber the effects of leaching.
- Results of outdoor tests disclosed that such competing processes can especially be assumed to occur in summer time. In autumn, higher amounts of target substances in the leachates were detected than in summer-time values.
- A model was developed that is able to predict the emissions found in the laboratory trials. This can be used for extrapolation purposes, as it makes allowance for the physico-chemical processes which are important to leaching behaviour.
- Analysis of outdoor data has shown that the amount of runoff is the most important factor influencing leaching. It can also be concluded that temperature and global radiation have a substantial impact on this. The total emissions detected in the outdoor tests can fluctuate considerably on a case by case basis and they depend on the prevailing weather conditions.
- There is a systematic difference between the results from lab tests and the findings from outdoor trials involving vertically aligned test specimens. The cumulative emissions determined in the lab are substantially higher than those for outdoors. Hence, the lab trials can be regarded as the worst-case scenario.
- The development of proper methods for obtaining reliable leaching data that can be used for risk assessments still poses a challenge. Preferably, experts drawn from all over Europe should act on this in concert.

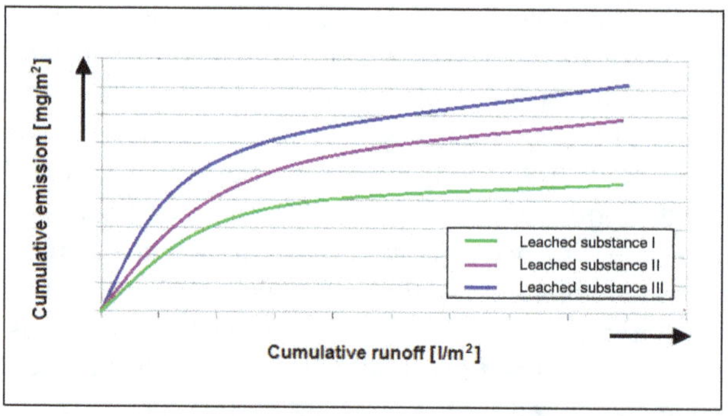

Figure 3.6: Schematic delineation of cumulative emissions per square metre versus cumulative runoff per square meter

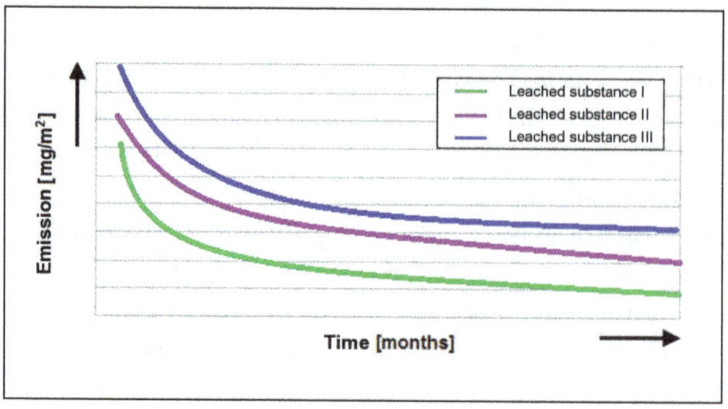

Figure 3.7: Schematic delineation of emissions versus time

A typical chart of outdoor leaching results is shown schematically in Figure 3.6, which presents cumulative emissions per square metre versus cumulative runoff per square meter. The blue curve (leached substance III) represents the worst leaching behaviour while the green curve (leached substance I) indicates the best leaching resistance of three test substances in a simplified test setup.

It can also be seen that the emissions often tend towards a limit value, i.e. leaching of a given substrate from a given matrix is relatively high at the outset, but typically decreases substantially over time due to the decline in concentration of the target substance in the uppermost layer of a coating. For further leaching, the target substance needs to be transported from the lower levels of the coating layer to the uppermost layer, a process that is reported to be mainly controlled by diffusion processes[272]. This behaviour can be more clearly illustrated by comparing the emission values over time periods, as shown schematically in Figure 3.7.

Due consideration of this effect is an important factor in the environmental risk assessment in order to avoid substantial overestimation of the leaching behaviour and consequently over-assessment of the potential risk of a given microbicide to the environment, especially bearing in mind that coatings often have double-digit service lives.

The results from the report described above[272] also reveal that a suitable transfer function for extrapolating results from lab trials or from outdoor tests to the real field still does not exist. Given that a large number of parameters influence the leaching behaviour of substances from a given coating material (such as matrix properties, physico-chemical properties of the target substances, test method, sample size, sample preparation, sample alignment, weather conditions, temperature, (UV) radiation), some scepticism might be allowed as to whether only the one transfer function can really be identified. It seems more likely that an entire set of transfer functions needs to be developed in order that all the various materials and all possible scenarios in the real field may be covered.

Equation 3.1

$$\text{Predicated emissions} \overset{?}{=} \begin{bmatrix} f(a_i, b_i, c_i, \ldots) \\ f(a_j, b_j, c_j, \ldots) \\ f(a_k, b_k, c_k, \ldots) \\ \vdots \end{bmatrix}$$

where: a, b, c = environmental parameters
i, j, k = matrix and active substance-related parameters

3.2 Optimisation of dosage

Given the many influencing factors, preliminary laboratory trials must be conducted of the microbial susceptibility of a product to be protected before a reliable statement about the necessary preservation measures can be made (see also Chapter 4).

The proper dosage of an in-can microbicide depends on various factors, especially the nature and sensitivity of the target product, the pH, the temperature, the initial microbial content, the extent of expected contact with germs and the envisaged shelf-life. Hygiene measures during production, storage and transport support optimisation efforts.

One way to determine the necessary dosage in the wet state is to carry out so-called in-can challenge tests (Section 4.3). In these, several samples of the same product under test are treated with different microbicidal products in varying concentrations. Different sets of the treated samples (for control purposes, each includes a blank sample without microbicide) are separately contaminated with either a specified bacterial, fungal, or yeast inoculum and then incubated for several days at a temperature specified in the method employed. This sample-inoculation procedure is repeated several times at weekly intervals. After each interval, all samples under test are subjected to a germ count. More details are provided in Section 4.3. A specimen test result is shown in Figure 3.8.

In this example, different specimens of the product under test are treated separately with two different in-can microbicide formulations containing comparable amounts of total active substance. The samples were inoculated five times with a bacterial suspension and, after each inoculation cycle, their germ counts were evaluated twice (after three and seven days) and expressed as the number of colony forming units (cfu) (see Section 4.2). Clearly, in this model case, in-can preservative I provides excellent protection from the outset of the test, even during four inoculation cycles; it is only after the fifth inoculation that the microbicidal performance starts to decrease (increase in bacterial count). In-can preservative II failed to successfully suppress the bacterial count throughout the test period, and so this formulation is unsuitable in this specific case. The simultaneous use of different concentrations of preservative I and II in the in-can challenge test can yield the optimum concentration, i.e. the lowest amount of active(s) that will produce satisfactory protection. It goes without saying that the same procedure needs to be carried out on further sets of samples with a view to assessing

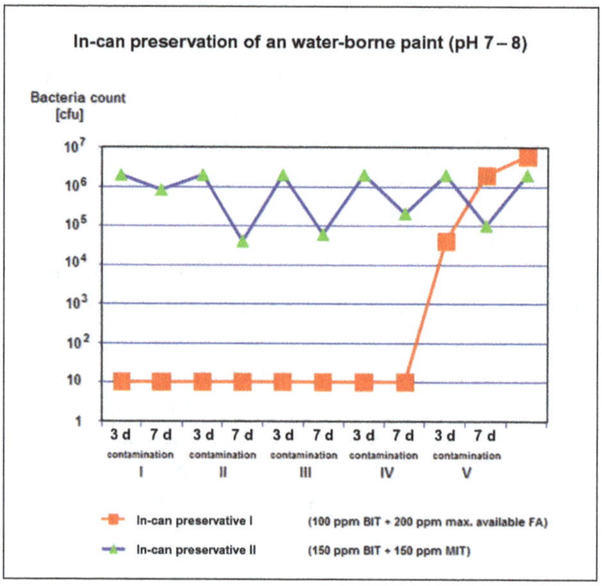

Figure 3.8: Specimen test result of an in-can challenge test on a water-borne paint

Figure 3.9: Result of an agar diffusion test against fungi;
top left: (1 % FF) edges covered marginally with mould (pass);
top right: (0.8 % FF) 25 to 50 % coverage with mould (fail);
bottom left: (0.6 % FF) fully covered with mould (fail);
bottom right: (0.4 % FF) fully covered with mould (fail)

the effectiveness of the microbicidal formulation at combating corresponding fungi and yeast inocula. In general, only that in-can microbicide which, at a given concentration in the product to be protected, successfully controls bacteria, fungi and yeasts can be recommended without any qualms for protection purposes.

Dosages of film preservatives can be optimised by employing the so-called agar diffusion (AD) test (see also Section 4.4). In particular, the following European Standards are available:

- EN 15457:2014 (Paints and varnishes – Laboratory method for testing the efficacy of film preservatives in a coating against fungi).
- EN 15458:2014 (Paints and varnishes – Laboratory method for testing the efficacy of film preservatives in a coating against algae).

The principles of the two are very similar. Samples of the liquid coating material under test are treated separately with different film preservative formulations in various concentrations. The samples are then applied to a substrate, conditioned according to the specifications and subsequently placed on jelled agar prepared in Petri dishes. Afterwards, the specimens (in triplicate) are inoculated with either a standard algal or a standard fungal suspension and incubated under the conditions detailed in the respective method. Finally, the efficacy of the film preservative is rated for the intensity of the fungal/algal growth on the coated surface of the test sample after incubation in accordance with a specified evaluation scheme. Figure 3.9 illustrates an exemplary outcome of such an AD test involving fungal inoculation of samples treated with 0.4 %, 0.6 %, 0.8 % and 1 % (wet/wet) of a fungicide formulation (FF) containing 4 % OIT and 8 % zinc pyrithione (representing a range of 160 to 400 ppm OIT and 320 to 800 ppm ZnP in the coating product).

3.3 Formulation aspects

As a basic rule, the way in which a microbicidal formulation is incorporated into a coating product to be protected depends on whether the intention is to provide in-can protection or film preservation.

In-can microbicides are ideally added directly to the feed water in order that process hygiene may be improved from the outset of a production sequence. However, if temperatures higher than 60 °C or a pH higher than 10 are expected, the addition of microbicidal formulations should be shifted to downstream production steps where conditions are less challenging. Furthermore, for reliable, uniform protection, suitable steps must be taken to ensure that the actives are homogeneously distributed in the product to be protected (e.g. stirring, agitating, circulating). Aqueous product preparations naturally freeze at low temperatures and should therefore be stored under frost-free conditions. Microbicidal products should be checked for incompatibility with other raw materials (e.g. sensitive polymer emulsions) before use.

Dry-film preservatives are best added (with stirring) towards the end of a production sequence, as this avoids the risk of degradation/decomposition in the presence of elevated temperatures, higher pH values or shear forces. Here, again, homogeneous distribution in the target product is essential for achieving uniform, reliable action. Film preservatives should also be checked for incompatibility with other raw materials in the coating formulation. Furthermore, dry-film preservatives should not be diluted with solvents or mixed with other chemicals prior to use.

3.4 Remedial surface treatment

Before redecoration starts on building façades, contaminated surfaces must be subjected to remedial work so that the microorganisms can be eradicated to the fullest extent possible. Unless such works are undertaken, there will always be the risk that some of the germs will survive the re-coating procedure and consequently continue their destructive action beneath the coating layer.

Fungal and/or algal infestations on exterior surfaces should be thoroughly cleaned with the aid of a pressure washer. Any plants or suckers of climbing plants should be removed mechanically or by flame.

Contaminants on interior surfaces should be wet-cleaned. In any case, it is important to observe all legal provisions and requirements. Additionally, it is strongly recommended that suitable protective clothing (overall, gloves, protective goggles) be worn during cleaning.

Figure 3.10: Remedial treatment of an infested external staircase; top left and right: initial situation; centre-left: cleaning with pressure washer; centre-right and bottom left: application of substrate pre-treatment agent either by brush or spray bottle; bottom right: final result

Residual amounts of infestation with fungi, algae or lichens should be thoroughly treated with a substrate pre-treatment agent, such as an aqueous solution containing 1 to 2 % alkyldimethylbenzylammonium chloride (ADBAC, CAS No.: 8001-54-5, see also Table 1.2 and Table 1.3). After the cleaned surfaces have been allowed to dry (approx. 24 hours), the new paint/coating can be applied. Figure 3.10 illustrates the remedial treatment of an external staircase.

3.5 New developments in the field of material protection

The placing on the market of biocidal actives and derived products is strictly regulated under European law[126] and this affects the biocides market in a number of ways. The registration effort required on the part of industry is complex, time-consuming and cost intensive. Starting from scratch, it would take an investment of roughly several million euros to get an active substance registered under BPR – this figure takes into account fees, study expenses as well as personnel and labour costs. It is not hard to understand that such investments will only pay off if an appropriate return on investment can be obtained in an acceptable timeframe.

As a result, several biocides are already being withdrawn from some applications and traditional chemistries are being replaced by other products. It therefore remains challenging from an industry perspective to strike the right balance between ecology and economics. One of the key targets is to prolong the service life of a substrate at a reasonable price. Sophisticated formulation technology is needed for improving, e.g., the leaching behaviour or the stability of a biocidal product in a given matrix.

3.5.1 Slow-release technology

Slow-release technology can effectively keep dry-film preservatives right where they are actually needed: on the coating surface. A number of coatings products have come on the market that claim to have slow-release properties and described as being 'encapsulated', 'embedded', 'protected', 'shielded', 'slow-release', 'controlled release' and the like.

For the encapsulation of microbicidal actives, a set of different methods has become available[278]. The desired release profile and the physico-chemical properties of the active substance to be encapsulated are the key parameters governing the choice of production method. Comprehensive information on slow-release activities can be found in the literature[278–282], which additionally provides a huge number of secondary references.

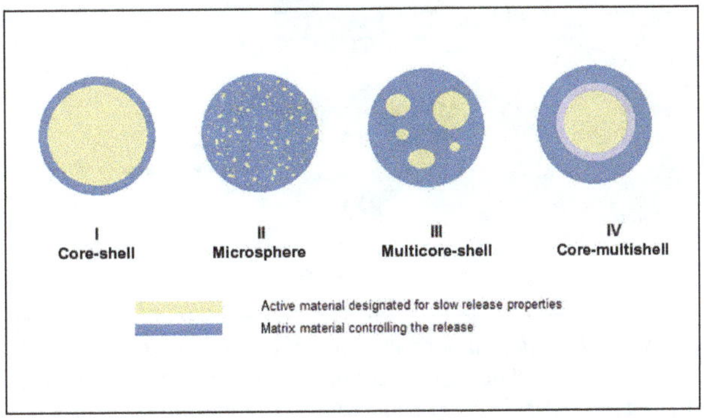

Figure 3.11: Different possible morphologies for slow-release particles

It should be noted that the morphology of a slow-release particle is not necessarily limited to a 'regular' capsule in the commonly understood meaning of the word, e.g. shaped like a cherry

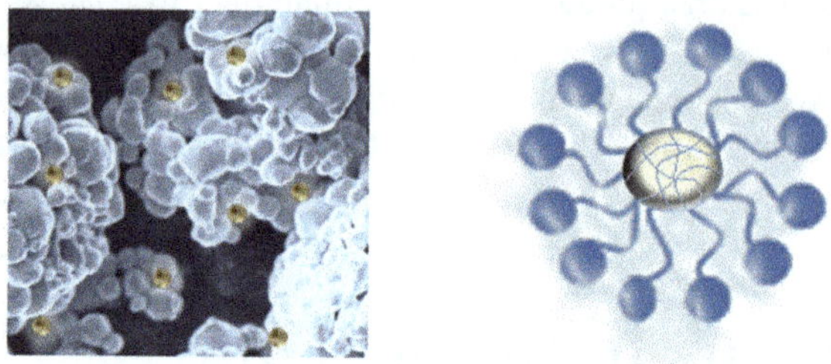

Figure 3.12: Schematic illustration of further possible slow-release techniques; left: by adsorption onto inert carrier particles; right: by encasement in micelle-like structures.

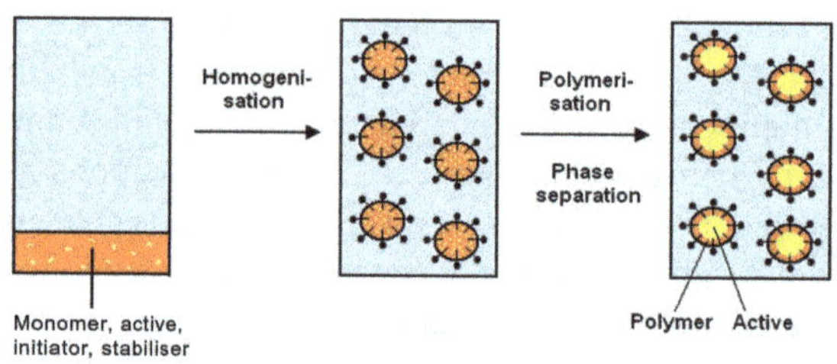

Figure 3.13: Example of a polymerisation process for creating slow-release particles

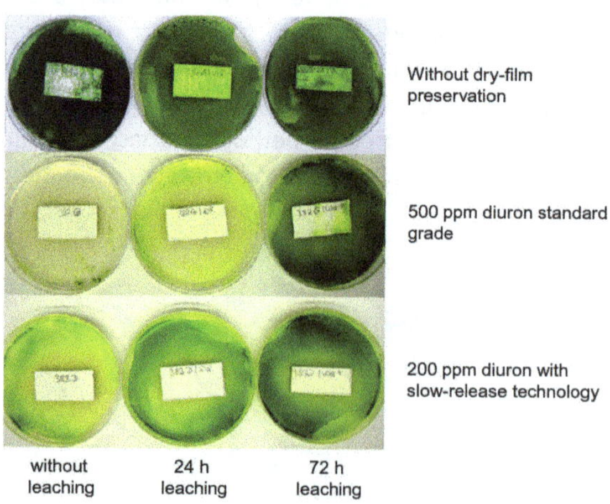

Figure 3.14: Algicidal performance of a slow-release formulation for diuron in a standard exterior emulsion paint; at the top: control without biocide; in the middle: diuron standard grade; at the bottom: slow-release formulation of diuron

pit. It is also possible for the active in question to be more or less homogeneously distributed in a polymeric matrix or to form multiple cores in a given microsphere[279]. Figure 3.11 schematically illustrates the various morphologies for controlled-release particles that are described in the literature[279].

Alternatively, it is theoretically possible to conceive of a microbicidal active formulation that provides slow-release by simply adsorbing the active substance onto an inert carrier material that has a relatively low desorption rate for this compound, as shown schematically in Figure 3.12 (left). A candidate carrier material here would be (modified) silica-based solids which could be deployed in a manner akin to that of fillers in coatings. Moreover, a slow-release effect could conceivably be achieved by encasing active substances in micelle-like structures, as schematically shown in Figure 3.12 (right); but in this case, at any rate, the structures would have to remain intact both when the coating is applied to the substrate and while it is curing afterwards.

The decisive point from an application perspective is whether the slow-release particle can withstand the storage conditions in a wet paint, e.g. throughout its shelf life, or whether the controlled-release actives are lixiviated into the liquid medium during storage, especially bearing in mind that a coating formulation typically contains a couple of additives, such as wetting and dispersing agents, surfactants and the like which could in theory interact with the matrix of the slow-release particle. As a result, the desired slow-release effect would not be maintained in the long term.

The ultimate slow-release particle would be an intelligent capsule or a cage-like structure featuring a 'switch' that would open the protective barrier only when the latter comes into contact with the excretion products (EPS) of microorganisms (see also Section 1.1.2). The microbicidal active would therefore only be released when it is actually needed and would otherwise remain encased. Were this to be realised, release of microbicidal actives into the environment via leaching processes would be substantially suppressed (see also Section 3.1).

One way to synthesise slow-release materials is illustrated in Figure 3.13[281]. In this case, the carrier medium for the polymerisation process is water. The monomers, which ultimately constitute the polymer matrix and which are virtually insoluble in water, are dispersed together with the biocidal active. The initiator is soluble in the monomer material, i.e. polymerisation occurs inside the monomer droplet. The droplets are usually stabilised by means of a protective colloid. At the start of polymerisation, the liquid is an emulsion, which is gradually transformed into a suspension. Due to increasing incompatibility between the growing polymer and the core material, phase separation takes place and the hydrophobic active becomes encapsulated in the polymer.

Standard agar diffusion tests (see also Chapter 4) to evaluate a slow-release formulation containing diuron against a standard grade reveal that comparable film protection against algal infestation can be achieved with much less active substance (Figure 3.14). In this case, only one-third to one-half of the original dosage is needed to produce satisfactory results[281]. Conversely, using the same primary dosage would substantially extend the service life of a coating treated with this slow-release preservative.

What is more, the corresponding EC50 value (algae) of diuron in this formulation was substantially improved by a factor of approx. 23. This corresponds to vastly superior behaviour in the environmental compartment compared with standard grades, without compromise to the biocidal performance at the intended point of action[281].

3.5.2 New actives

It was mentioned in the introduction to Section 3.5 that the launch of a microbicidal active that is previously unavailable in the European market requires time-consuming and cost-intensive registration effort on the part of industry. It is therefore not surprising that there are not very many new actives currently undergoing the evaluation process under BPR[126]. Table 3.3 shows a list of new actives (as per 4 May 2017) for product types 7 (film preservatives), 8 (wood preservatives), 9 (fibre, leather, rubber and polymerised materials preservatives) and 10 (construction material preservatives) for which a new application for approval has been submitted[180].

In particular, the active substances azoxystrobin (CAS No. 131860-33-8) and fludioxonil (CAS No. 131341-86-1) are of some interest (see also Figure 3.15), not least due to their classification and labelling.

In the context of harmonised classification and labelling (CLH) throughout the European Union, a dossier was submitted by the Danish Competent Authority for the new active fludioxonil ('new' in the sense of the BPR)[283]. It proposes that the future entry in Annex VI of the CLP Regulation[196] should not contain a classification related to human health hazard, like the classification and labelling of thiabendazole (see Section 2.2.1.2) which also is not classified in this regard.

Azoxystrobin as such is classified as being toxic by inhalation according to the harmonised classification and labelling indicated in Annex VI of the CLP Regulation[196].

In all cases, it is classified for aquatic toxicity, but this is also true of the vast majority of film preservatives described in the previous sections. With regard to the latter, however, almost all have received substantial classification and labelling in respect of human health hazards, starting from sensitising properties, to toxicity by the oral, dermal or inhalation route to the point of suspected carcinogenicity or to reproduction toxicity.

This makes these new actives very attractive for use as film preservatives in specialised indoor applications.

Azoxystrobin belongs to the strobilurin class of fungicides. The fungicidal activity of this chemical family was discovered during research into special mushrooms which are found in European forests and which possess a remarkable ability to defend themselves against natural adversaries; they have consequently come under scrutiny from scientists[284–287]. The defensive mechanism is based on the release of chemical substances called strobilurin A and oudemansin A (Figure 3.16).

Further activities aimed at developing analogues and derivatives of the above-mentioned compounds identified azoxystrobin as belonging to the most active and stable representatives that also suppress spore germination and mycelial growth[287].

Due to their novel mode of action, strobilurin compounds have emerged as important agricultural products, with the first examples being sold in 1996 and now holding a substantial market share in the plant protection business[284–287].

Azoxystrobin works by inhibiting mitochondrial respiration[288, 289]. It binds to a specific protein which plays a major role in converting chemical energy from nutrients into a form which the fungal cells can utilise. As a result of protein binding, the fungi run out of energy and die. Azoxystrobin can be employed against all classes of fungal pathogens, including species responsible for exuding foul smells, staining and generally biodeterioration (e.g. *Cladosporium spp.* or *Alternaria spp.*), because the mode of action targets fungi universally.

In terms of the evaluation process under BPR[126], the active is expected to be discussed by the relevant Biocidal Products Committee (BPC) in the course of 2017. The BPC prepares the opinions of ECHA on several BPR processes (including applications for approval and renewal of approval for active substances). The final decisions are then taken by the European Commission.

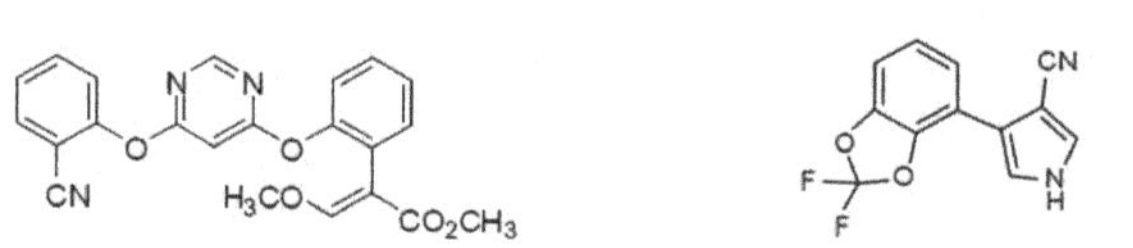

Azoxystrobin
(Methyl (2E)-2-(2-{[6-(2-cyanophenoxy) pyrimidin-4-yl]oxy}phenyl)-3-methoxyacrylate)
CAS n° 131860-33-8

Fludioxonil
(4-(2,2-difluoro-1,3-benzodioxol-4-yl)-1H-pyrrole-3-carbonitrile
CAS n° 131341-86-1

Figure 3.15: Molecular formulas of azoxystrobin and fludioxonil

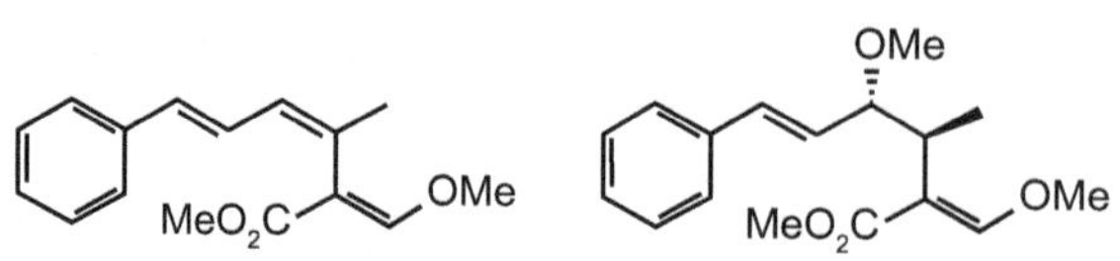

Strobilurin A

Oudemansin A

Figure 3.16: Molecular formulas of strobilurin A and oudemansin A

Table 3.3: Applications for approval of new actives for use in product types 7, 8, 9, 10

Active substance name	CAS n°	PT	Country of entity	Inclusion reason	Inclusion date
Allyl isothiocyanate	57-06-7	9	Ireland	new active	29 May, 2015
Sodium metabisulfite	7681-57-4	9	Netherlands	new active	10 Feb, 2015
Pythium oligandrum, Chromista – Stramenopila	not allocated	10	Czech Republic	new active	24 Sep, 2014
2-octyl-2H-isothiazol-3-one (OIT)	26530-20-1	8	Germany	new active	24 Sep, 2014
N-(trichloromethylthio) phthalimide (Folpet)	133-07-3	7	Germany	new active	24 Sep, 2014
N-(trichloromethylthio) phthalimide (Folpet)	133-07-3	9	Germany	new active	24 Sep, 2014
Silver nitrate	7761-88-8	7	Germany	new active	17 Aug, 2015
Silver nitrate	7761-88-8	9	Germany	new active	17 Aug, 2015
Granulated copper	not allocated	8	Germany	new active	24 Sep, 2014
Penflufen	494793-67-8	8	Germany	new active	13 Jan, 2016
Fludioxonil	131341-86-1	7	Germany	new active	27 Feb, 2015
Fludioxonil	131341-86-1	9	Germany	new active	27 Feb., 2015
Fludioxonil	131341-86-1	10	Germany	new active	27 Feb, 2015
Azoxystrobin	131860-33-8	7	Germany	new active	23 Nov, 2015
Azoxystrobin	131860-33-8	9	Germany	new active	23 Nov, 2015
Azoxystrobin	131860-33-8	10	Germany	new active	23 Nov, 2015
Reaction mass of chloromethyl hexyl cyano-carbonodithioimidate and dihexyl cyanocarbono-dithioimidate	not allocated	9	Belgium	new active	23 Mar, 2017

The mode of action of fludioxonil is attributed to inhibition of the osmotic signal pathway, leading to impairment of fungal spore germination and prevention of mycelial growth. This mode of action is instantaneous.

Fludioxonil provides innate activity against several *Penicillium spp.*, *Alternaria spp.*, *Stachybotrys chartarum* as well as some wood-decaying fungi (e.g. *Conophora puteana*, *Gloeophyllum trabeum* and *Sydowia pythiophila*), and others.

The BPC opinion on the approval of the active substance fludioxonil in product type 7, 9 and 10 was adopted on 2 March 2017. The opinion is published on the ECHA webpage.

3.6 Microbicides based on silver compounds

This silver-based class of chemical microbicidal actives will only be discussed briefly here. Such compounds have been used for a fairly long time, e.g. in the context of hygiene coatings that claim to exert an antibacterial surface effect (e.g. in hospitals, surgeries, on doorknobs, and handrails). They have also been employed in woven fabrics and textiles, such as under-

Table 3.4: Applications for approval of silver-based microbicides (as per 4 May, 2016)

Substance Name	CAS n°	Product type	Legal act	Evaluating Competent-Authority	Approval status
Disilver oxide	20667-12-3	11	Decision 2014/227/EU	SE	not approved
Reaction mass of titanium dioxide and silver chloride	not allocated	1, 2, 6, 7, 9, 10, 11		SE	all PT's under review
Silver	7440-22-4	2, 4, 5, 11		SE	all PT's under review
Silver adsorbed on silicon dioxide (as a nanomaterial in the form of a stable aggregate with primary particles in the nanoscale)	1167997-68-3	9		SE	under review
Silver chloride	7783-90-6	3	Decision 2014/227/EU	SE	not approved
Silver chloride	7783-90-6	4	Decision 2014/227/EU	SE	not approved
Silver chloride	7783-90-6	5	Decision 2014/227/EU	SE	not approved
Silver chloride	7783-90-6	13	Decision 2014/227/EU	SE	not approved
Silver nitrate	7761-88-8	1, 2, 3, 4, 5, 7, 9, 11, 12		SE	all PT's under review
Silver phosphate glass	308069-39-8	2, 7, 9		SE	all PT's under review
Silver sodium hydrogen zirconium phosphate	265647-11-8	1, 2, 4, 7, 9		SE	all PT's under review
Silver zeolite	not allocated	2, 4, 5, 7, 9		SE	all PT's under review
Silver copper zeolite	130328-19-7	2, 4, 5, 7, 9		SE	all PT's under review
Silver zinc zeolite	130328-20-0	2, 4, 5, 7, 9		SE	all PT's under review

PT 1 - Human hygiene
PT 2 - Disinfectants and algaecides not intended for direct application to humans or animals
PT 3 - Veterinary hygiene
PT 4 - Food and feed area
PT 5 - Drinking water
PT 6 - Preservatives for products during storage
PT 7 - Film preservatives
PT 9 - Fibre, leather, rubber and polymerised materials preservatives
PT 10 - Construction material preservatives
PT 11 - Preservatives for liquid-cooling and processing systems
PT 12 - Slimicides
PT 13 - Working or cutting fluid preservatives

wear, and bed mattresses that claim antimicrobial effects as well. Further fields of use for which an application for approval under BPR has been submitted are indicated in Table 3.4.

There have been some quite controversial discussions in the market about the merits and downsides of using microbicidal silver compounds in coatings. Claimed advantages (hygiene products for human welfare, disease control, etc.) have been juxtaposed against the disadvantages, such as relatively weak fungicidal effects, discolouring issues, possible development of germ resistance and the relatively high price.

In the interim, five substance/PT combinations based on silver have failed to receive approval[290] under the BPR evaluation process (see also Table 3.4), with the result that the future of this chemical family under BPR would seem to be in doubt, at the very least.

4 Microbiological and application test methods

In order to have a microbicidal active substance or a corresponding microbicidal product based on this active approved under European legislation[126], it is necessary to demonstrate that the active substance has an innate ability to produce an effect on a representative target organism. This requirement is to intended to prevent unhelpful chemical substances from being placed on the market purely for commercial/marketing reasons. In cases, too, where the substance under evaluation is usually combined with other active substances in a biocidal product, it is a requirement to show that the former possesses innate activity on its own. However, it is not necessary to demonstrate efficacy against all of the target organisms as early as the active substance approval stage, as additional target species might be added subsequently in the product authorisation stage.

Although a great many different test methods for evaluating the efficacy of microbicides are currently available in the global markets, they are not harmonised for many applications, with the consequence that the test results obtained are not necessarily comparable or even inconsistent. For this reason, a set of test methods that should find broad acceptance is presented in the following sections.

4.1 Minimum inhibitory concentration

The determination of the minimum inhibitory concentration (MIC) is a well-known laboratory method for determining the innate antimicrobial properties of a biocide[291–293]. The method is designed to identify the lowest concentration level of an active needed to inhibit or to prevent visible growth of the test species under controlled laboratory conditions. Similarly, the minimum microbicidal concentration (MMC) is that amount of active which kills the specified microorganisms. The method is particularly applicable to bacteria, moulds, yeasts and algae. The corresponding effective dosages are usually given in mg/l or in ppm. The resulting values can then be used to estimate the probably effective dosage of an antimicrobial agent in a product to be protected (e.g. a paint in the wet state) and to identify efficacy gaps which might be closed by combining the active substance under test with other microbicidal substances. One example, already mentioned in Section 2.1.4, concerns a combination of BIT and bronopol for in-can protection purposes. BIT possesses significant efficacy gaps against *Pseudomonas spp.* which can effectively be closed by adding, for instance, bronopol to the microbicidal formulation.

A typical test procedure for bacteria consists in inoculating different standardised test samples containing liquid culture medium as well as different concentrations of an antimicrobial agent. After incubation for a specified time period (e.g. 24 h/3 d/5 d), the bacterial growth is assessed and the MIC value is recorded[291].

4.2 Determination of germ count

A determination of the germ count in a product under test is *inter alia* important in the field of plant hygiene (see also Section 2.3) and as a precursor to performing in-can challenge testing as part of dosage optimisation (see also Sections 3.2 and 4.3).

The samples taken during a plant hygiene audit (e.g. process water, raw materials, swabs or final product) need to be checked for their content of viable microorganisms and appropriate measures must be initiated if the results disclose a significant microbial burden. In-can challenge testing for dosage optimisation is only useful when the samples under test are more or less germ-free or have only a low germ count, as otherwise falsified results will be obtained. The reason is that a microbicide under test would already have been consumed to a certain extent at the outset of the test procedure due to the initial microbial burden. The subsequent challenge with specified amounts of germs might therefore not be passed, even though the originally intended antimicrobial dosage would have been effective enough. Consequently, the microbicidal performance of an active substance formulation would not be a true representation in this case.

A number of methods for determining germ counts are available[294, 295] and the choice of a particular procedure depends *inter alia* on the target species under test and the respective areas of application (e.g. medicine, food control, material protection).

A typical approach in the coatings area is the spread plate procedure that allows the number of actively growing/dividing germ cells in a test specimen to be estimated. This method is illustrated below for the enumeration of the bacteria count in a liquid test sample. Similar test protocols are also available for assessing the microbial burden in test samples for fungi and yeasts[294, 295].

A set of Petri dishes is partially filled with sterilised liquid agar containing nutrients for the growth of bacteria, often along with reagents that resist the growth of non-target organisms (some recent methods also include a fluorescent agent for automated evaluation of the test results). Subsequent cooling to ambient conditions lets the agar solidify so that the dishes are now ready to be inoculated ("plated") with the test material that supposedly contains bacteria (see Figure 4.1). The test material is diluted as necessary and a small aliquot is transferred from a sterile pipette to an agar plate. With the help of a sterilised loop, the sample material is spread over the agar surface by a special streaking technique in which the loop is worked back and forth across the plate.

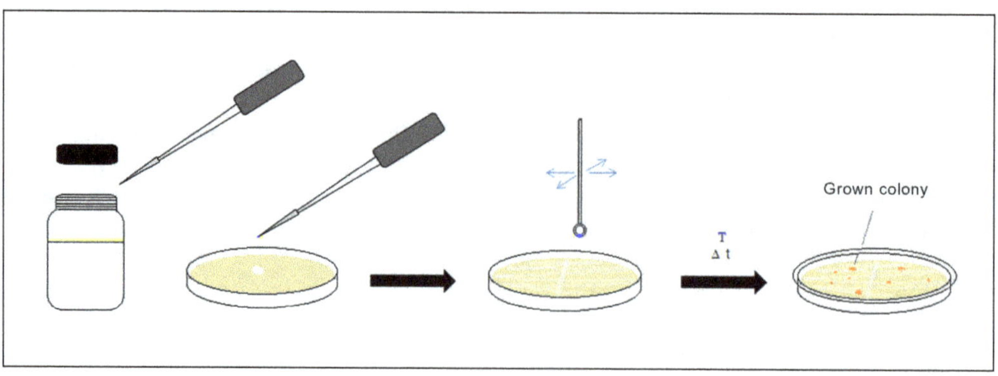

Figure 4.1: Schematic illustration of the spread plate procedure

Subsequently, the plates are covered with the Petri dish lids and allowed to stand for several minutes before they are inverted for incubation at a temperature (e.g. 30 °C) and a time (e.g. 2 days) specified in the method employed. Inversion of the dishes is designed to reduce the undesirable risk of contamination by airborne particles settling on the agar surface and to prevent condensation from potentially interfering water dripping from the inside of the lid onto the culture. At the end of the incubation period, the plates are evaluated by counting all colonies that can be detected (e.g. with the aid of a plate counter equipped with a light source and a magnifying glass) and the number of bacteria is then calculated for the original sample. Here, it is assumed that each viable bacterium in the original test specimen generates individual colonies on the agar plate under the given test conditions, provided that the test procedure is carried out properly. The evaluation procedure just described is more suitable than counting all the bacteria under a microscope; the counting process for which may be complicated by a substantial number of dead cells that are of no interest. Finally, the determined values of colony forming units (cfu) are usually subjected to a rating scheme as indicated in Table 4.1.

Table 4.1: Exemplary rating scheme for germ count

Rating n°	cfu/ml	Description
0	–	no growth
1	10^0 to 10^1	traces of growth
2	10^1 to 10^2	slight growth
3	10^2 to 10^3	moderate growth
4	> 10^3	heavy growth

An alternative procedure is the so-called pour plate method which is often employed when the analysis is aimed at bacterial species that grow poorly in air. In this case, a diluted test specimen supposedly containing bacteria is mixed with melted agar and then that mixture is poured into a Petri dish. After solidification and subsequent incubation, the grown colonies are counted and the number of viable cells is then calculated.

4.3 In-can challenge test

As discussed in Section 3.2, the performance of an antimicrobial formulation designed to protect the wet state of coating products depends on various factors, especially the nature and the sensitivity of the target product, the pH, the temperature, the extent of expected contact with germs and the envisaged shelf-life. The level of performance is also contingent on proper handling during production, storage and transport.

Several methodologies are available for evaluating the performance of antimicrobial formulations designed to protect coating products in the wet state[296–300]. However, they vary in respect of test parameters, such as inoculum type, incubation conditions, and sampling/evaluation schedule. Consequently, the results obtained might not necessarily be consistent. This is all the more important bearing in mind that tests involving living matter are inherently fraught with uncertainty. Microbial strains grown in the lab, for instance, might be less viable than analogous strains collected from real-life environmental compartments.

One of the most prominent methods for assessing in-can protection capabilities in the coatings area was published by the International Biodeterioration Research Group (IBRG) [297–299], an organisation consisting of scientists, biocide manufacturers, academic institutions and test facilities with the main focus on test method development and investigating biodeterioration issues. Many aspects of in-can protection were reviewed with the aim of choosing the best options from a set of test parameters. The basic procedure is described in the following for the bacterial challenge of a paint product (see also Figure 4.2).

Several samples of the same paint without any microbicides are treated with different microbicidal products in various concentrations. Different sets of the treated samples (plus a blank sample without microbicide for control purposes in each case) are contaminated separately with a defined bacterial inoculum as follows. Each bacterial species of the test inoculum is subcultured separately in liquid medium. A mixed suspension of different bacterial species is immediately prepared just before the inoculation procedure (at least 10^8/ml for each bacterium species). 1 ml of the prepared inoculum is then added to 50 g paint. The inoculation procedure is repeated three times at weekly intervals. Four hours, twenty-four hours, forty-eight hours and seven days after each inoculation cycle, the bacterial count is determined by the streak plate method (see Section 4.2). The results are then rated according to the scheme in Table 4.1. All tests are run in duplicate.

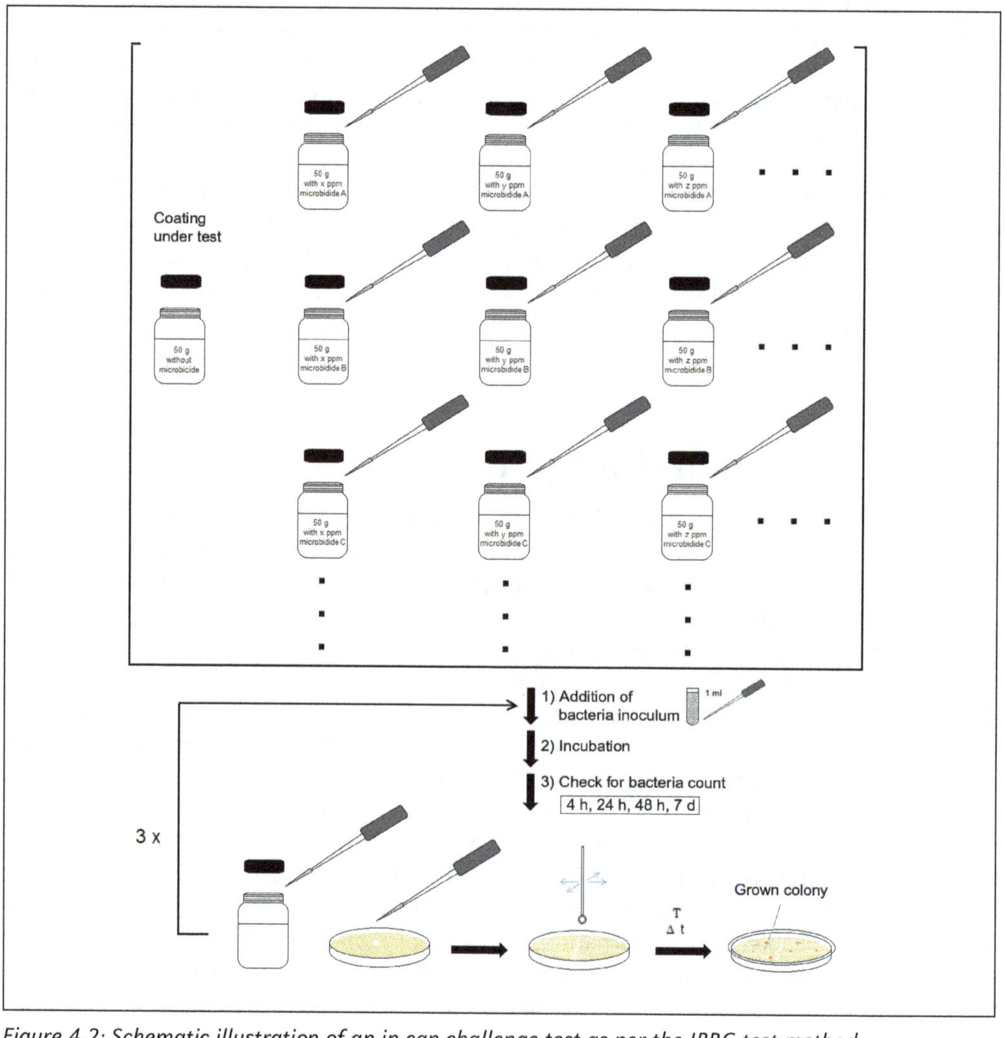

Figure 4.2: Schematic illustration of an in-can challenge test as per the IBRG test method

4.4 Agar diffusion test

As already discussed in Section 3.2, the efficacy of film preservatives can be estimated in laboratory trials by performing an agar diffusion (AD) test. The following European Standards are specifically available for coatings:

- EN 15457:2014 (Paints and varnishes – Laboratory method for testing the efficacy of film preservatives in a coating against fungi).
- EN 15458:2014 (Paints and varnishes – Laboratory method for testing the efficacy of film preservatives in a coating against algae)

The principle is very similar in both cases and so both are described below. Basically, this entails making a semi-quantitative comparison of coatings, with and without film preservatives (see also Figure 4.3).

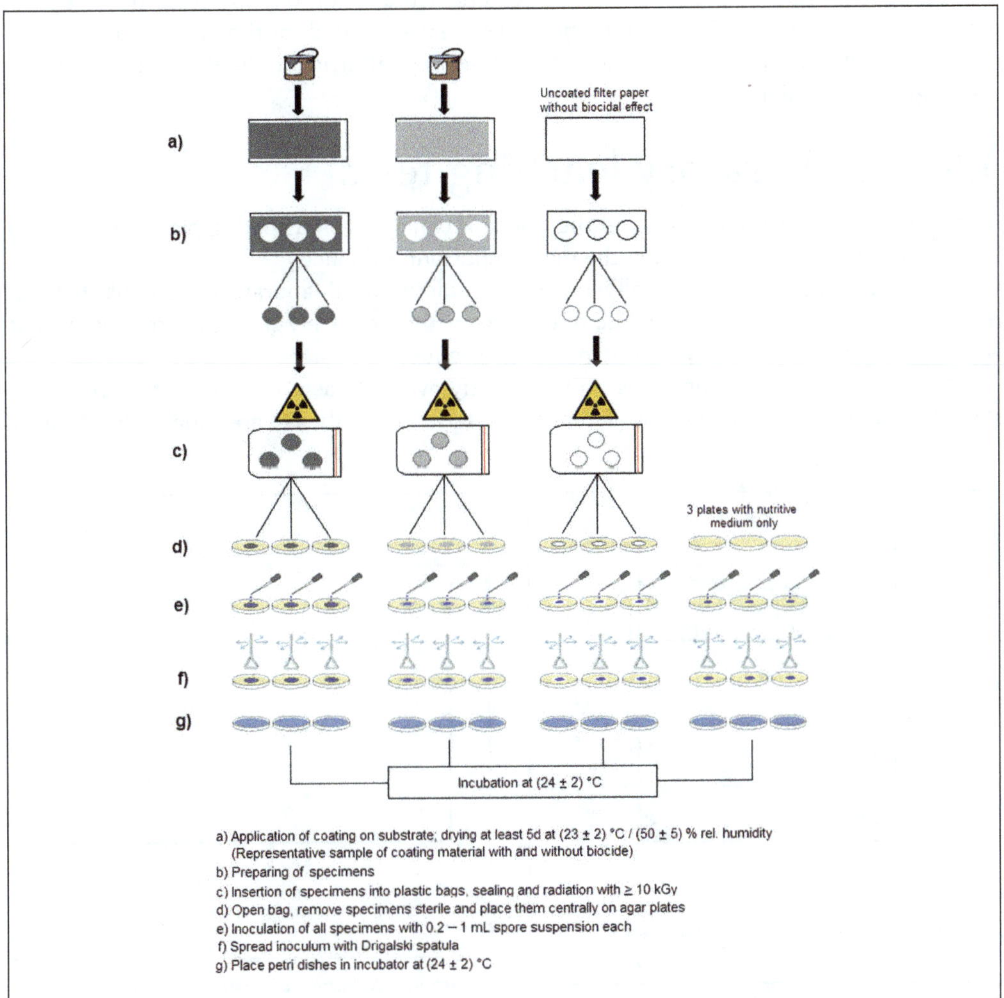

Figure 4.3: Schematic illustration of an AD test against fungi

The liquid coating material under test is treated with different film preservative formulations in various concentrations. The test samples are then applied to an appropriate substrate. A strip of filter paper without microbicidal effect is usually employed for this purpose, but if the filter strip cannot be kept flat due to undulation, other substrates may be used instead of filter paper, provided that they do not inhibit fungal/algal growth. The coated test substrates are conditioned in horizontal position for at least 5 days at (23 ± 2) °C and (50 ± 5) % relative humidity. After the conditioning period, three specimens of the coated test substrates are prepared with a diameter of 55 mm each, and are then sealed in plastic bags and sterilised with gamma radiation (≥ 10 kGy). Test specimens without film preservative (in triplicate) and test specimens of the uncoated substrate (also in triplicate) are included as controls for each test series.

Subsequently, a set of sterile Petri dishes is prepared with the respective culture medium (malt agar for the AD test against fungi; algal nutritive agar for the AD test against algae). The test specimens are placed on the corresponding jelled agar surfaces, inoculated with the defined algal/fungal suspension and incubated under the conditions specified in the respective method. Finally, the efficacy of the film preservative is rated for the intensity of the fungal/algal growth on the coated surface of the specimen after incubation and in accordance with a given evaluation scheme.

4.5 Laboratory leaching tests

It has been shown in Section 3.1 that the leaching behaviour of a coating plays a major role in its service life, with runoff being the most important influence.

Industry is therefore continually looking for standardised laboratory methods that can reliably predict the leaching behaviour of coatings within relatively short time frames, as that will enable it to improve the performance of its products.

Although such a laboratory method is already available as a European standard[301], it should be noted that the cumulative emissions determined in the lab are usually significantly

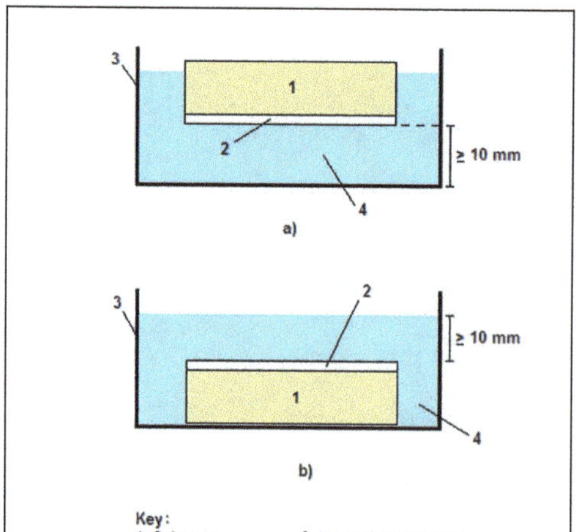

Figure 4.4: Schematic test setup of the laboratory immersion test as per EN 16105:2011

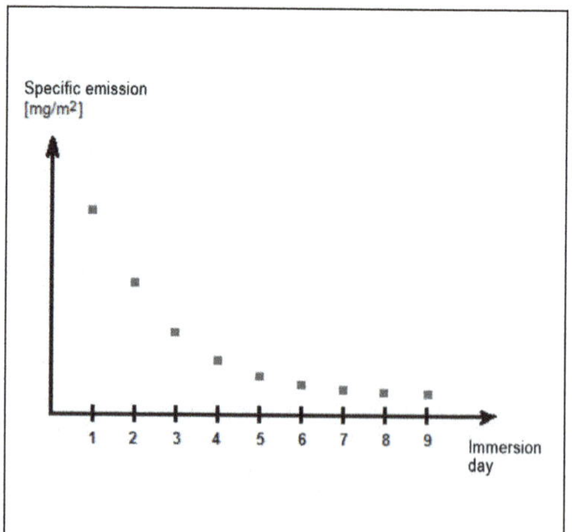

Figure 4.5: Schematic illustration of the typical outcome of the laboratory immersion test as per EN 16105:2011

higher than those detected in outdoor tests (see Section 4.6) and therefore such laboratory trials seem to substantially overestimate the leaching behaviour of a coating under test. This was recently confirmed by a comprehensive report published by the German Federal Environmental Agency[272] regards the emissions of film preservatives into the environment (see also Section 3.1). The European Standard itself addresses this issue in its definition of the scope of the laboratory method[301]:

- "This European Standard specifies a laboratory method to determine the leaching behaviour of substances from coatings into water over defined time intervals. **The release of substances from coatings under natural conditions cannot be determined with this method.**"

The laboratory procedure for determining the release of substances from coatings via intermittent contact with water according to the European standard[301] is described briefly below (see also Figure 4.4).

Coated specimens are periodically immersed in water under specified conditions. In between the dipping cycles, the samples are allowed to dry. One dipping cycle consists of 1 h immersion, 4 h drying and again 1 h immersion. Nine dipping cycles (e.g. on days 1, 3, 5, 8, 10, 12, 15, 17 and 19) are performed. The leaching waters for each immersion day are combined and analysed for the identity and quantity of leached compounds.
The dipping container should be large enough to completely expose the coated surfaces to water. Arithmetically speaking, one square metre of coating surface under test should be exposed to 25 litres of water; in other words, test specimens of 100 cm^2 area are dipped in 250 ml water.

The substrate used for applying the coating layer needs to have a homogeneous planar surface and should be inert (e.g. XPS, EPS). Absorbent substrates need to be sealed on the reverse side and around the edges to avoid penetration of water. The mass of each test specimen is recorded before each immersion day and the test samples are allowed to dry for at least 42 hours between two immersion cycles. Finally, the analytical results of each target substance, expressed in mg/L (if detectable), are converted to specific emissions (mg/m^2) and contrasted against the respective immersion day. A schematic illustration of a typical outcome of such a laboratory trial is presented in Figure 4.5.

4.6 Semi-field leaching trials

One possible test setup for outdoor leaching tests is based on a procedure known as NT Build 509 (2005) that was originally developed for testing the leaching behaviour of treated wood for Use Class 3 ('outdoor above ground')[302]. This procedure can be adapted analogously as follows for testing specimens covered with exterior coatings:

Suitable panels are prepared, vertically assembled and exposed outdoors to the normal environment. The panels can be oriented in a way that they are exposed to the maximum expected driving rain (worst case). According to NT Build 509, three replicas for each test assembly are required. However, to keep study costs down, duplicate exposure seems justified. This issue is currently under discussion at European level because clarification is needed in case it is intended that the tests will be consulted for environmental risk assessments in the context of registering actives under European law (see also Section 5).

The leachate is collected and monitored by chemical analyses of the active substances. Laboratory glass flask or plastic jars with no impurities that might influence the active substance analysis can be used for this purpose (2 to 3 litre container might be needed for an expected annual rain load of 700 mm). The quantity of rain, the wind direction and the wind speed at the test site need to be monitored by a weather station. Long-term leaching assessment usually entails exposing the test racks for at least one year, but more likely for two years. Runoff samples might be taken and analysed after each rain event in the first months of the test period. Later on, the runoffs will be collected and then combined after a certain time period to streamline the analytical effort. Figure 4.6 illustrates a typical test panel assembly in a semi-field leaching test.

Figure 4.6: Test panel assembly in a semi-field leaching test in Germany (western orientation)

4.7 International standards

Table 4.2 indicates a selection of test methods that are employed in addition to those already described in Section 4.

Table 4.2: Selection of international test methods employed in the coatings area

Method	Title	PT
ASTM D2574-06	Standard test method for resistance of emulsion paints in the container to attack by microorganisms	6
ASTM D4783-01e1	Standard test methods for resistance of adhesive preparations in container to attack by bacteria, yeast, and fungi	6
SABS 1102 (1987)	Bacterial efficacy of biocides used in water-based emulsion paints	6
ASTM G21-96(2002)	Standard practice for determining resistance of synthetic polymeric materials to fungi	7
ASTM D3273-00(2005)	Standard test method for resistance to growth of mold on the surface of interior coatings in an environmental chamber	7
ASTM D3456-86(2002)	Standard practice for determining by exterior exposure tests the susceptibility of paint films to microbiological attack	7
ASTM D5589-97(2002)	Standard test method for determining the resistance of paint films and related coatings to algal defacement	7
ASTM D5590-00(2005)	Standard test method for determining the resistance of paint films and related coatings to fungal defacement by accelerated four-week agar plate assay	7
ASTM E1428-99(2004)	Standard test method for evaluating the performance of antimicrobials in or on polymeric solids against staining by *streptoverticillium reticulum* (a pink stain organism)	7
ASTM G29-96(2002)	Standard practice for determining algal resistance of plastic films	7
ASTM D1006-93	Standard recommended practice for conducting exterior exposure tests of paints on wood	7
AATCC 30 - 1999	Antifungal activity, assessment on textile materials: mildew and rot resistance of textile materials	9
ASTM E2471-05	Standard test method for using seeded-agar for the screening assessment of antimicrobial activity In carpets	9
OECD (ENV/JM/BCID(2007)5)	Guidance document on the evaluation of the efficacy of antimicrobial treated articles with claims for external effects	9
CTBA-BIO-E 002	Field ageing test on treated masonry	10
ASTM WK8681	Standard test method for resistance to mold growth on interior coated building products in an environmental chamber	10
ASTM D4939-89 (2007)	Standard test method for subjecting marine antifouling coating to biofouling and fluid shear forces in natural seawater	21
ASTM D3623-78a(2004)	Standard method for testing antifouling panels in shallow submergence	21
ASTM D5479-94(2007)	Standard practice for testing biofouling resistance of marine coatings partially immersed	21

5 Legislative aspects

Within the European Union, the legislative aspect plays a decisive role when it comes to judiciously choosing an active from a set of actives for antimicrobial use, because the following simplified principle applies:

No registration, no market!

In a first step, applicants need to obtain the approval for an active ingredient. In a second step, authorisation for biocidal products based on such actives needs be granted as well before they can be placed on the market[1]. In addition, the underlying complex process is linked to specific application areas that are clustered in so-called product types (PTs). These application aspects need to be reflected in the relevant human health and environmental risk assessments as part of the BPR evaluation process. An active substance that, for instance, has been approved for film preservation purposes (PT 7) cannot be used for preserving construction materials when no application for approval for use in PT 10 has been made, even though the use pattern of the corresponding biocidal product could be very similar in both cases.

Outside Europe, e.g. in the USA, Canada and Mexico, comparable and extensive registration procedures are also in place that require a comprehensive set of data to be submitted for the purpose of obtaining biocidal product registration so that the product can be placed on the relevant market.

The requirements related to the registration/authorisation procedures are manifold and it would far exceed the scope of this book to cover all aspects in detail, since the sheer scale of information, provisions and the complexity of the topic, with its plethora of alterations, adaptations, interpretations and cross-references, would preclude a simple listing. In addition, it is not unlikely that some details correct at the time of going to print could be outdated by the time the book is read (e.g. due to the enactment of new regulations or amendments to existing regulations that trigger a new factual and legal position on certain topics). Therefore, only a selection of principal points can be highlighted here. This information is provided in good faith and to the best of the author's knowledge but comes without warranty and does not release readers from the obligation of doing their own verification. Readers seeking further information should consult the references given in the following sections. Any such 'volunteers' should be aware that such an endeavour is not for the faint-hearted.

[1] Exemptions from these rules exist and are described on following pages.

Frank Sauer: Microbicides in Coatings

5.1 Biocidal Product Legislation (BPR)

5.1.1 General aspects of the authorisation process

The legal framework for placing biocidal products on the European market is the current Biocidal Products Regulation (EU) No. 528/2012 ('BPR')[126] which is the 'successor' to the Biocidal Products Directive 98/8/EC ('BPD')[181]. The BPR was adopted on 22 May 2012 and entered into force on 17 July 2012, the twentieth day following its publication in the Official Journal of the European Union. The date on which the BPR first applied was 1 September 2013. As a 'Regulation' the BPR has legal binding force in all EU countries, unlike the former BPD which, as a 'Directive', had to be transposed into the national laws of each Member State.

The implementation of the BPD was the first approach towards harmonising the regulations of biocidal actives and products based thereon throughout the European Union which, up until then, varied considerably across the Member States. In this context, the BPD was also intended to provide an adequate level of protection for humans, animals and the environment, an ambition that has also found its way into the BPR. Further objectives of the latter are to improve the functioning of the European market for biocidal products and to remove a number of deficiencies that were identified during the implementation of the BPD.

5.1.1.1 Authorisation of biocidal active substances

For substances that, as an integral part of biocidal products, were already on the market on 14 May 2000 ('existing' biocidal active substances), the European Commission set up a review programme under the BPD that continues under the BPR. This programme covers identified existing biocidal actives for which a notification was accepted under Commission Regulation (EC) No. 1451/2007, Annex II(303). This regulation was repealed and replaced by the Commission Delegated Regulation (EU) No. 1062/2014(182), which has adapted the detailed rules for the Review Programme to the provisions of the BPR. Biocidal products containing active substances which are included in the above-mentioned Review Programme but which have not yet been approved for a given product type can still be placed on the European market and used under the terms of the transitional provisions as laid down in Article 89-95 of the BPR. One aspect in this connection is that the relevant Member State, acting in accordance with its national rules, may allow the marketing of such products in its territory until the last active substance in a biocidal product formulation has been approved for the respective product type, and what is more, for up to three years after the last approval date (shorter time frames may apply in the event of non-approval).

For biocidal products not covered by the scope of the BPD but which are now subject to the BPR (Article 93) and which were available on the market on or before 1 September 2013, applications for approval of the relevant active substance in those products had to be submitted to the European Chemicals Agency (ECHA) before 1 September 2016. Otherwise, the product must be removed from the market by 1 September 2017. Two groups of substances are affected by this deadline[308]:

- active substances used in food-contact materials that provide antimicrobial surface action (e.g. antimicrobial substances incorporated into plastic chopping boards)
- *in situ*-generated active substances that originate from a different starting material – the 'precursor' – at the place of use. In this context, the relevant precursors did not fall under the scope of the BPD because either they were not placed on the market, or no claim

was made in respect of a biocidal purpose. Under the BPR, however, the potential risks of both the precursor and the substance it generates are to be evaluated (e.g. ozone generated from oxygen from the air).

Commission Delegated Regulation (EU) No. 1062/2014 also adapts the procedures for the dossier evaluation to align them with the BPR processes for new active substances and provides a defined role for the European Chemicals Agency (ECHA), referred to below as the 'Agency'. The approval of active substances takes place at Union level and the subsequent authorisation of the biocidal products at Member State level. Such product authorisations can be extended to other Member States by mutual recognition. A dedicated IT platform, the Register for Biocidal Products (R4BP 3), is used for submitting applications, exchanging data and information between the applicant, ECHA, Member State competent authorities (MSCA's) and the European Commission.

The evaluation of an active substance dossier is, in the first instance, made by a selected EU Member State which acts as the evaluating Competent Authority (eCA). The latter provides a draft Competent Authority Report (CAR) which is subsequently discussed at Union level in the Biocidal Products Committee Working Groups (BPC WG) which support the Biocidal Products Committee (BPC) with the preparation of its opinions and contribute to the harmonisation of risk assessment under the BPR. Finally, the BPC adopts an opinion on the approval or non-approval of the relevant active substance. The legally binding decision is made by the European Commission.

The main steps involved in active substance approval under BPR are illustrated in Figure 5.1.

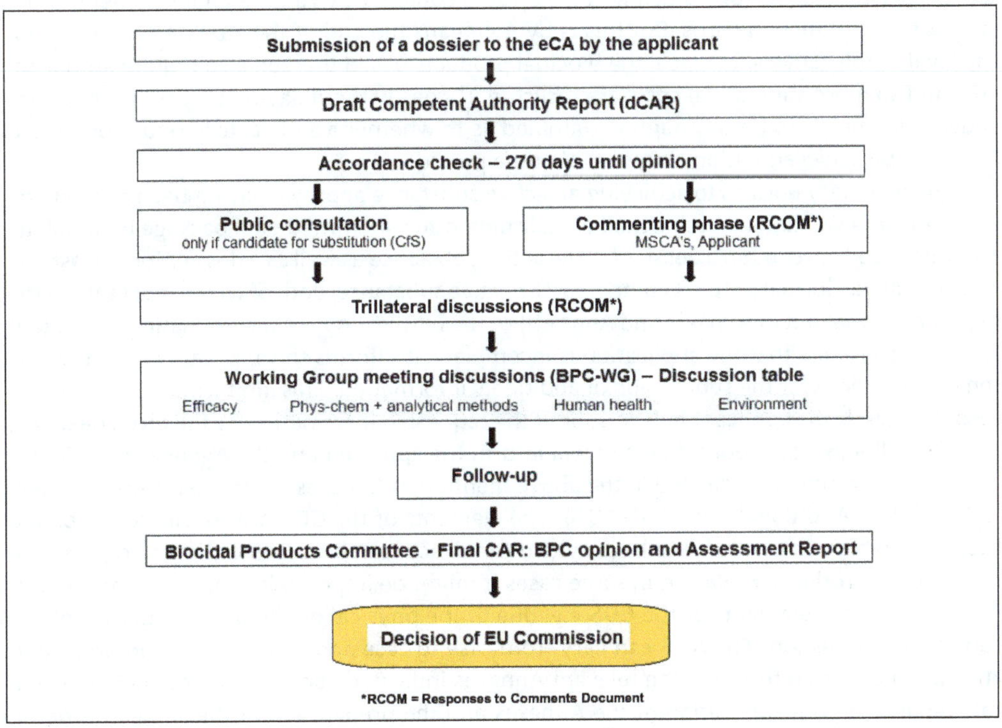

Figure 5.1: Main steps involved in active substance approval under BPR

Furthermore, Commission Delegated Regulation (EU) No. 1062/2014(182) supplies clarification on certain issues related to participants in the Review Programme and it introduces the possibility of adding substance/PT combinations to the Review Programme under certain conditions. Finalisation of the work programme is now expected by 2024(182). Besides, it specifies time limits for process of evaluating an active substance. After receipt of an application for approval, the Agency informs the applicant (also referred to as the 'participant') of the fees payable under Commission Implementing Regulation (EU) No. 564/2013(304). The standard approval fee for a new active substance under Article 7(2) of the BPR, for instance, is EUR 120,000 for the first product type. An extra fee of EUR 40,000 is charged per additional PT for which the applicant seeks approval. Fee reductions may be granted to small and medium-sized enterprises (SMEs), provided they are established in the European Union. This financial support is not granted if the active substance is a candidate for substitution (see below). Should the participant fail to pay the fee within 30 days, the application will be rejected. Otherwise, the Agency will accept the application and inform both the participant and the designated evaluating Competent Authority (eCA) accordingly, indicating the date of its acceptance and its unique identification code.

Further fees in an amount similar to those described above arise via the eCA, which charges applicants directly for its evaluation work, under Article 80(2) and (3) of the BPR. Here, again, the applicant will be informed by the eCA of the fees payable within 30 days of the Agency's receipt of the application. The latter can be rejected by the eCA if the participant fails to settle the account in time.

Additional costs for the preparation of the data package and handling expenses (e.g. for project management, consultant etc.) need to be considered as well. To obtain approval of an active substance under BPR starting from scratch could easily require an investment in the range of several million euros. But this is by no means the end of the story. After successful approval of the active substance, the biocidal products based thereon also require authorisation, and so more substantial costs are incurred. It goes without saying that such a business undertaking needs to be thoroughly examined as to whether a satisfactory return on investment can be achieved within a reasonable time period.

The monetary hurdles to achieving an active substance approval having been cleared, the eCA initiates a validation check aimed establishing if the submitted data package is complete. The latter needs to contain a dossier for the active substance as well as a dossier for at least one representative biocidal product containing the active substance, both of which must satisfy the requirements set out in Annex II and Annex III of the BPR[126]. Where the evaluating competent authority considers that the application is incomplete, it informs the applicant and sets a reasonable time limit for the submission of additional information. This limit will normally not exceed 90 days. If the applicant fails to submit the requested information by the given deadline, the eCA will reject the application and will inform the applicant and the Agency accordingly.

The data elements indicated in the above-mentioned Annexes comprise a 'Core Data Set' (CDS) and an 'Additional Data Set' (ADS). The elements of the CDS are considered to be the basic data which usually need to be provided for all active substances and for all representative biocidal products. However, in some cases it might be impossible or unnecessary to provide specific data elements of the CDS e.g. due to the physical or chemical properties of the substance in question. The ADS can vary from case to case and so the readers interested in knowing more are referred to the relevant Annexes indicated above. Further detailed technical guidance on the application of the Annexes and the preparation of the respective dossiers is available on the Agency's website[305].

The studies conducted or referenced need to be described in full and explicitly, and should include all the methods employed. The various test protocols must be compliant with those laid down in the REACH Regulation[306] and specified in detail by Council Regulation (EC) No. 440/2008[307]. Where test data exist that were generated before 1 September 2013 (the date on which the BPR first became applicable – see above) by methods that do not meet the requirements of Commission Regulation (EC) No. 440/2008 and its amendments, the relevant Member State Competent Authority will decide on a case-by-case basis whether such data is adequate for the purpose of approving an active substance under BPR. In particular, the need to minimise testing on vertebrates must be taken into account, among other factors.

The submitted data must be relevant and of sufficient quality to meet the requirements of the BPR. In addition, it is necessary to demonstrate that the active substance under test was the same as the substance for which the application has been filed.

The applicant is also obliged to initiate a pre-submission consultation with the designated evaluating Competent Authority at which the proposed information requirements may be discussed, especially the prevention of tests on vertebrates (see also the obligation set down in Article 62(1),(2) of the BPR regarding forced data sharing).

The application for approval of an active substance will be evaluated by the Competent Authority, which may, where relevant, include proposals to adapt data requirements. A corresponding draft assessment report and the conclusions of its evaluation will then be sent to the Agency. Where several applicants support the same substance/product-type combination, only one assessment report will be drafted. The eCA must observe the following time limits for accomplishing this (whichever is the later):

- either within 365 days after acceptance of completeness for the substance/product-type combination in question
- or within the time limits provided for under Annex III of Commission Delegated Regulation (EU) No. 1062/2014 (see also Table 2.1).

In this context, the applicant has 30 days to make a written comment on the assessment report and its associated conclusions. Those comments are to be given due consideration by the eCA when it is finalising its evaluation and before it is submitted to the Agency. The eCA may also ask the applicant to submit within a specified time limit additional information which is regarded as necessary for carrying out the evaluation. In that event, the Agency is informed accordingly. Furthermore, the time limit for finalising the evaluation process by the eCA will be suspended from the date of issue of the request until the date the information is received ('stopping the clock'). The suspension may not exceed 365 days, where the additional information relates to concerns not addressed under the former BPD, and 180 days in all other cases. Exceptions may apply in extraordinary circumstances or due to the nature of the data requested.

Identified concerns for human health, animal health or the environment that might result from the use of biocidal products containing the active substance under evaluation will be documented in the conclusions of the eCA.

Under conditions specified in Article 6(7) of the Commission Delegated Regulation (EU) No. 1062/2014, the eCA will submit upon finalisation of its hazard evaluation, without undue delay and no later than at the time of submission of the assessment report, a proposal to the Agency for harmonised classification and labelling ('CLH dossier') of the substances in question (and, where appropriate, specific concentration limits or M-factors, or a proposal for a revision thereof) pursuant to Article 37(1) of the 'CLP Regulation'[196].

Upon receipt of the eCA report, the Agency will prepare and submit to the Commission an opinion on the approval or non-approval of the relevant substance/product-type combination. This opinion is prepared and adopted beforehand by the Biocidal Products Committee (BPC) which is associated with the Agency. The following time-limits apply in respect of starting the preparation of the opinion by the Agency (whichever is the later):
- either within three months of receipt of the eCA report
- or within the time limits provided for by Annex III of Commission Delegated Regulation (EU) No. 1062/2014 (see also Table 2.1).

In general, the BPC provides support related to the following BPR processes:
- applications for approval and renewal of approval of active substances
- review of approval of active substances
- eligibility for the simplified authorisation procedure (see also Annex I of the BPR)
- identification of active substances which are candidates for substitution (see below)
- applications for Union authorisation of biocidal products (see below) and for renewal, cancellation and amendments of Union authorisations, except where the applications are related to administrative changes
- scientific and technical matters concerning mutual recognition
- preparing an opinion on any other questions that may arise from the operation of the BPR relating to risks to human or animal health or the environment or to technical guidance, at the request of the Commission or of the Member States

Member States are each entitled to appoint one member to the BPC for a renewable term of three years. They may also appoint an alternative member. Applicants may participate in BPC discussions. The respective agendas are published no later than 21 days before such a meeting.

The BPC is in turn supported by Working Groups (WG) who assist with the preparation of its opinions. The Working Groups also contribute to harmonising risk assessment under the BPR, carry out scientific and technical peer reviews and consider other relevant scientific and technical issues within the scope of the BPR. All Working Groups report to the BPC subject to its rules of procedure. Applicants may also participate in BPC Working Group (BPC WG) discussions. The respective agendas are published no later than 21 days before such a meeting.

Once preparation of the opinion has started, it must be submitted within 270 days to the Commission, which takes the final decision. On receipt of the opinion of the Agency, the Commission shall either
- adopt an Implementing Regulation (IR) that an active substance is approved and indicating the specific conditions thereof, the date of approval as well as the expiry date of the approval; or
- adopt an Implementing Decision that an active substance is not approved in cases where
    - the conditions for approval under Article 4(1) of the BPR are not met; or
    - where applicable, the conditions for the exemption from the 'Exclusion Criteria' set out in Article 5(2) of the BPR are not fulfilled (see below); or
    - where the requisite information and data have not been submitted within the prescribed period.

Approved active substances will be included in a respective Union list that is kept up to date by the Commission, which also makes the list electronically available to the public.

In this regard, the relevant active substance will typically be approved for an initial period not exceeding 10 years, although some restrictions may apply. These are described below.

Article 5(1) of the BPR states that active substances can normally not be approved when they meet one or more of the following criteria ('exclusion criteria'):

- active substances which have been classified or which meet the criteria to be classified as
  - carcinogenic: category 1A or 1B; or
  - mutagenic: category 1A or 1B; or
  - toxic for reproduction: category 1A or 1B.
- active substances which meet the criteria for being
  - PBT (persistent, bioaccumulative and toxic); or
  - vPvB (very persistent and very bioaccumulative).
- active substances which are
  - identified as having endocrine-disrupting properties that may cause adverse effects in humans; or
  - considered as having endocrine-disrupting properties that may cause adverse effects in humans.
    (On 15 June 2016, the Commission adopted a communication presenting the latest state of play on the file and the way forward[309]. However, further controversial discussions took place on European level. A comprehensive updated overview regards this topic is given on the website of the European Commission[310].)

Nevertheless, substances falling under these exclusion criteria of the BPR may be approved if it is shown that at least one of the following conditions laid down in Article 5(2) of the BPR is met:

a. "the risk to humans, animals or the environment from exposure to the active substance in a biocidal product, under realistic worst case conditions of use, is negligible..."; or
b. "it is shown by evidence that the active substance is essential to prevent or control a serious danger to human health, animal health or the environment"; or
c. "not approving the active substance would have a disproportionate negative impact on society when compared with the risk to human health, animal health or the environment arising from the use of the substance".

The decision by the authorities in this regard needs to consider whether suitable and sufficient alternative substances or technologies might be available. An active substance that falls under the exclusion criteria paragraphs of the BPR may only be approved for an initial period not exceeding five years.

A further restriction regarding authorising active substances is laid down in Article 10 of the BPR that deals with the so-called 'candidates for substitution'. A substance is considered a candidate for substitution if any of the following conditions are met:

- it meets at least one of the exclusion criteria according to Article 5(1) of the BPR but may be approved in line with Article 5(2) of this Regulation (see above)
- it meets the criteria to be classified as a respiratory sensitiser in accordance with the CLP Regulation[196] and its respective amendments
- its acceptable daily intake (ADI), acute reference dose (ARfD) or acceptable operator exposure level (AOEL), as appropriate, is significantly lower than those of the majority of approved active substances for the same product-type and use scenario
- it meets two of the criteria for being PBT (persistent, bioaccumulative and toxic) in accordance with Annex XIII to the REACH Regulation[306]

- there are reasons for concern related to the nature of the critical effects which, in combination with the use patterns, may lead to use that could still give cause for concern, even with very restrictive risk management measures (e.g. high potential of risk to groundwater)
- it contains a significant proportion of non-active isomers or impurities.

The Agency/BPC examines whether the active substance under evaluation fulfils any of the criteria for being a candidate for substitution and addresses the matter in its opinion that is sent to the Commission. Prior to this, information on potential candidates for substitution is made publicly available by the Agency during a period of no more than 60 days in which interested third parties may submit relevant input, including data on available substitutes. The Agency will give due consideration to the received information in its final opinion. Active substances considered to be candidates for substitution may only be approved for an initial period not exceeding seven years. In that event, they are identified as such in the relevant Implementing Regulation.

The provisions for the renewal of an approval of an active substance are set out in Articles 12 to 14 of the BPR. At least 550 days before the expiry date of the approval of an active substance, a respective application for renewal has to be submitted to the Agency, along with all relevant data (see Article 13 of the BPR). Where the substance has different expiry dates for different product types, the earliest expiry date is relevant for the above-mentioned submission deadline.

The renewal is only granted if the active substance still meets the relevant requirements laid down in the BPR. Where appropriate, the specific conditions for approval of an active substance can be amended by the Commission in the light of scientific and technical progress.

The period of validity of the renewal is usually 15 years for all product types to which the approval applies, unless a shorter period is specified in the respective Implementing Regulation of the Commission. The approval of a candidate for substitution may only be renewed for a period not exceeding seven years.

Where there are significant indications that the conditions for approval of an active substance under the relevant provisions in the BPR are no longer met, the Commission may at any time review such approvals for one or more product types (Article 15 of the BPR). This also applies to circumstances where Member States request such a review by the Commission due to significant concerns about the safety of biocidal products or treated articles containing the active substance in question. The initiation of a review process is made publicly available by the Commission and the respective applicant is given the opportunity to submit comments which will be given due consideration in the review process.

5.1.1.2 Authorisation of Biocidal Products

The general principles concerning the authorisation of biocidal products are quite similar in parts but some are even more complex than those for a biocidal active substance, and so only some of the crucial points are presented here. Readers interested in more information are referred to the relevant provisions laid down in the BPR (especially Articles 17 – 24, 29 – 34 and 41 – 46) and to the available notes and guidance documents on the ECHA website[311].

An authorisation may be granted for a single biocidal product or a biocidal product family (Article 17(3) of the BPR). The latter comprises a group of biocidal products for similar application purposes that contain active substances having the same specifications. The compositions of the relevant products in a family may vary only to a specified extent and under the proviso that the level of risk is not increased or the efficacy of the product is not reduced.

The variability in composition can either take the form of a reduction in the active substance concentration or a variation in the non-active substance contents. The substitution of non-active substances for other non-actives is also possible under the condition that they do not pose a greater risk.

All products within a biocidal product family are covered by a single authorisation granted under BPR. This concept is a further development of the so-called 'frame formulation' approach under BPD. The acceptable ranges for a biocidal product family will be established in the authorisation process by means of assessments and evaluations by relevant representative members. Where a new product is also covered by the established ranges, it needs only to be notified by the authorisation holder to the relevant authorities 30 days before its placing on the market. Notification is not even necessary if the variations only take the form of dyes, perfumes or pigments within existing permitted ranges.

Following the successful evaluation/authorisation of the active substances contained in a biocidal product, a subsequent product authorisation is necessary in those Member States where this product is intended to be marketed. In principle, there are three options for achieving a biocidal product (family) authorisation under BPR:

1. either at a national level by submitting a dossier to a selected Member State ('Reference Member State') and mutual recognition in other concerned Member States; or
2. in certain cases also at EU level ('Union Authorisation'), allowing for direct access to the entire EU market (which is new under BPR); or
3. through a simplified procedure for those products containing only active substances laid down in Annex I of the BPR.

In the first case, there are two further options. The process for mutual recognition in the concerned Member States can either be initiated **after** the first authorisation of a biocidal product has been granted in the Reference Member State ('mutual recognition in sequence'), or the application process can be started simultaneously in all chosen countries ('mutual recognition in parallel'). For this, the application for product authorisation is submitted to the Member State of choice, together with a list of other Member States where a national authorisation is sought. At the same time, the application procedure for mutual recognition in parallel in the other concerned Member States is started. The last option is faster and reduces the administrative burden.

A Union Authorisation of biocidal products is for the time being only possible for selected product types and subject to the provision that they must have similar conditions of use across the Union. Biocidal products containing active substances which meet the exclusion criteria under Article 5 of the BPR (see above) cannot be granted a Union Authorisation. Furthermore, product types 14, 15, 17, 20 and 21 are also exempted from this possibility.

Depending upon the product type, eligibility to apply for a Union Authorisation follows this phase-in scheme (Article 42 of the BPR):

1. from 1 September 2013 on: biocidal products containing one or more new active substances and biocidal products of product types 1, 3, 4, 5, 18 and 19
2. from 1 January 2017 on: biocidal products of product types 2, 6 and 13
3. from 1 January 2020 on: biocidal products of all remaining product types.

The Union Authorisation process starts with the submission of an application to the Agency, including a confirmation that the biocidal product would have similar conditions of use across the Union. The applicant also proposes a Member State Competent Authority (MSCA) that should evaluate the application. For this, the applicant needs to provide written confirmation that the Competent Authority agrees to do so. A Union Authorisation is valid throughout the Union, unless otherwise specified. It confers the same rights and obligations in each Member State as a national authorisation.

A further option for a product authorisation is a simplified procedure (Article 25 – 28 of the BPR) for a product that only contains certain active substances specified in Annex I of the BPR. Such products must not contain any substances of concern or any nanomaterials, they must be sufficiently effective for their purpose and their handling must not require protective equipment. The processing time for a simplified authorisation is much faster than for the other options described above. A further advantage is that relevant products can be made available in all European markets without the need for mutual recognition. Examples of actives meeting the criteria for a simplified procedure are substances authorised as food additives (e.g. lactic acid or sodium benzoate) and traditionally used substances of natural origin, such as lavender oil and peppermint oil.

5.1.2 Article 95: List of active substances and suppliers

The current status as to whether an active substance is currently under review, already approved or even non-approved by the European Authorities can readily be investigated on the corresponding ECHA website[305].

Moreover, the European Competent Authority regularly publishes a list of active substances and suppliers[180] on its website (also referred to as the 'Article 95 List'), including existing participants in the former BPD Review Programme.

As of 1 September 2015, it is illegal to place a biocidal product on the European market if the substance supplier or product supplier is not included in the Article 95 List for the relevant product type(s). Article 95 of the BPR aims to ensure equal treatment of all involved market players and fair allocation of the costs related to the substance approval process during the period when manufacturers or importers place this active substance on the market. This is different from the BPD, where companies could remain in the market even though they were not participating in the Review Programme ('free riders').

The following parties were placed automatically on the list and thus did not have to make a submission under Article 95 (BPR):

- participants in the Review Programme
- supporters of new active substances (submitters of a dossier under Article 11 of the BPD or under Article 7 of the BPR)
- submitters of third party dossiers (alternative active substance dossiers submitted as part of a product authorisation application).

Entities which are typically affected by Article 95 of the BPR and which therefore needed to make a submission to the Agency in time in order to maintain continued supply are:

- manufacturers of active substances in the Review Programme who were not participants in this connection
- importers of active substances (on their own or in biocidal products) in the Review Programme who were not participants in this regard
- manufacturers of new active substances who did not support their approval
- importers of new active substances (on their own or in biocidal products) who did not support their approval

- manufacturers of biocidal products containing active substance(s) from non-listed suppliers
- entities making available biocidal products on the market that contain active substance(s) from non-listed suppliers

5.1.3 Treated articles

According to Article 3(1)(l) of the BPR, a 'treated article' means any substance, mixture or article which has been treated with, or intentionally incorporates, one or more biocidal products.

As of 1 September 2013, treated articles may only contain such active substances (incorporated via a corresponding biocidal product) that either have already been approved under BPR or are still under evaluation for use in the relevant product type(s). However, a transitional period was granted until 1 September 2016 for those active substances which were not yet in the approval process. By that date, the relevant company had to submit a complete application dossier on such active substances, including data on the respective product type. As of 1 March 2017, treated articles that do not fulfil one of the above-mentioned criteria may no longer be placed on the European market. Where an active substance is not approved for the relevant product type, the respective treated article needs to be withdrawn from the market over a grace period of 180 days starting from the decision of non-approval on the active substance. This is different from the BPD, where articles imported from third countries could be treated with substances not approved in the EU.

When requested by consumers, suppliers must be ready to provide information about the biocidal treatment of treated articles within 45 days and free of charge. In addition, manufacturers and importers of treated articles need to ensure that products are properly labelled where this is a requirement of Regulation (EC) No. 1272/2008 on classification, labelling and packaging of substances and mixtures[196] and the Biocidal Products Regulation[126]. The relevant provisions of the latter are set out in Article 58(3). A treated article needs to be labelled when:

a. a claim is made by the manufacturer of that treated article regarding its biocidal properties; or
b. it is required in the conditions associated with the approval of the active substance(s) used for such treatment (via a respective biocidal product), having particular regard to the possibility of contact with humans or the release into the environment.

An example taken from practice of the corresponding wording of the conditions mentioned under case b) is given below in *italics:*
"The placing on the market of treated articles is subject to the following condition: The person responsible for the placing on the market of a treated article treated with or incorporating ... shall ensure that the label of that treated article provides the information listed in the second subparagraph of Article 58(3) of the Regulation (EU) No. 528/2012."

Where labelling is required under BPR, the following information needs to be provided on the label:

- a statement that the treated article incorporates biocidal products
- the biocidal property attributed to the treated article (where substantiated)
- the name of all active substances contained in the biocidal products
- where applicable, the name of all nanomaterials followed by the word 'nano' in brackets
- any relevant instructions for use, including any precautions to be taken.

The labelling needs to be clearly visible, easily legible and suitably durable. It may alternatively be printed on the packaging, on the instructions for use or on the warranty when the size or the function of the treated article does not allow for attachment of such a label. The required information must be given in the official language or languages of the Member State of introduction, unless that Member State provides otherwise. In special cases (e.g. for tailor-made products), the manufacturer may agree with the customer on other methods of providing the necessary information.

Although the above-mentioned conditions for labelling seem to be fairly clear, there often exists confusion in this context, especially regarding the use of the terms 'treated article' versus 'biocidal product' and 'biocidal property' versus '(primary) biocidal function'. The definition of the term 'treated article' has already been given above. A biocidal product according to Article 3(1)(a) of the BPR is:

- "any substance or mixture, in the form in which it is supplied to the user, consisting of, containing or generating one or more active substances, with the intention of destroying, deterring, rendering harmless, preventing the action of, or otherwise exerting a controlling effect on, any harmful organism by any means other than mere physical or mechanical action
- any substance or mixture, generated from substances or mixtures which do not themselves fall under the first indent, to be used with the intention of destroying, deterring, rendering harmless, preventing the action of, or otherwise exerting a controlling effect on, any harmful organism by any means other than mere physical or mechanical action."

A 'biocidal property' of a treated article means a characterising trait resulting from the biocidal product it incorporates with the intention to prevent the action of harmful organisms.

This term covers both the biocidal action on the treated article itself and the action conferring a 'biocidal function' on the treated article.

A treated article with a 'biocidal function' is an article which is not intended to protect the article itself or its original function, but to introduce an *additional purpose* which is biocidal.

A treated article without biocidal function can nevertheless have a biocidal property that provides protection from microbial decay and thus increases durability of the article itself (e.g. extension of the lifespan of façade paints or prevention of the deterioration of plasticised PVC). Conversely, every treated article with a biocidal function automatically has a biocidal property.

The term 'primary biocidal function' can be interpreted as a prominent property of the treated article in question that confers a biocidal function of first rank, importance, or value (compared with other functions of the treated article). This term is only used in Article 3(1)(a) of the BPR and is not specified further there. However, it is specified that a treated article with a primary biocidal function is considered a biocidal product (and as such will require product authorisation). Whether a biocidal function of a treated article is primary or not must be decided on a case-by-case basis. In this context, all individual properties and functions of the treated article as well as its intended uses should be taken into account. It is also important to thoroughly check the corresponding label claims in this regard, as misleading declarations may lead to unwanted labelling requirements.

In contrast to treated articles, substances or mixtures only need to have an intended biocidal function to fulfil the definition of a biocidal product (that triggers a necessary product authorisation), irrespective of whether the biocidal function is primary or not. Comprehensive guidance on this topic is given on the ECHA website[311].

5.2 Interrelationship of the BPR and other legislation

It is obvious from the previous sections that the BPR is not an isolated system of rules. There are numerous cross-references to and dependencies on other EU Regulations, especially
- Regulation (EC) No. 1272/2008 on classification, labelling and packaging of substances and mixtures – the 'CLP regulation'[196]
- Regulation (EC) No. 1907/2006 of the European Parliament and of the Council of 18 December 2006 concerning the Registration, Evaluation, Authorisation and Restriction of Chemicals – 'the REACH regulation'[306]
- Council Regulation (EC) No. 440/2008 of 30 May 2008 laying down test methods pursuant to Regulation (EC) No. 1907/2006[307]
- Commission Implementing Regulation (EU) No. 564/2013 of 18 June 2013 on the fees and charges payable to the European Chemicals Agency pursuant to Regulation (EU) No. 528/2012[304]
- Commission Delegated Regulation (EU) No. 1062/2014 of 4 August 2014 on the work programme for the systematic examination of all existing active substances contained in biocidal products referred to in Regulation (EU) No. 528/2012 – the 'Review Programme Regulation'[182]
- Directive 98/8/EC of the European Parliament and of the Council of 16 February 1998 concerning the placing of biocidal products on the market – the 'BPD'[181].

Besides these, there are Commission Delegated Regulations, Commission Implementing Regulations and Commission Implementing Decisions which are not explicitly mentioned here, but which are of relevance in this context. Not to be forgotten either are the respective amendments and adaptations to scientific and technical progress as well as a vast array of notes and guidance documents that provide additional information. It goes without saying that it would be quite a challenge to retain an overview and not lose sight of the central theme. Three other Regulations will also be mentioned here which do not directly deal with biocidal products, but which could indirectly have an impact in this regard:
- Regulation (EC) No. 1107/2009 of the European Parliament and of the Council of 21 October 2009 concerning the placing of plant protection products on the market and repealing Council Directives 79/117/EEC and 91/414/EEC[312]
- Regulation (EC) No. 1223/2009 of the European Parliament and of the Council of 30 November 2009 on cosmetic products[185]
- Regulation (EU) No. 305/2011 of the European Parliament and of the Council of 9 March 2011 laying down harmonised conditions for the marketing of construction products and repealing Council Directive 89/106/EEC[313].

Some active substances that fall under the regime of the BPR might also be used in the field of plant protection or in the cosmetics area, where they need a separate authorisation. Where concern substantiated by evidence arises in the course of an evaluation process for an active substance under these regimes, it is likely that these findings will sooner or later be considered under the BPR, and vice versa.

Further relationships with other legal provisions are possible, e.g. in the context of the European Ecolabel, the preservation of materials used for toys, and materials with direct or indirect food contact. In all these cases, it is essential to check whether additional requirements need to be met.

6 Summary and outlook

Prolonging the service life of substrates at a reasonable price level without triggering an unacceptable risk to the consumers and to the environment is a key demand.
Biocidal products can effectively protect various substrates, thereby conserving natural resources and supporting the sustainability mindset. In addition, these products might help to minimise risks to human health caused by pathogenic germs.

Almost all biocides are regarded as dangerous substances, a fact which is hardly surprising as their purpose is to control the growth of microbial life. But when handled with care and bearing in mind that microbicides are still essential for protecting billions of square metres of surfaces, this class of substances makes an important contribution not only to the preservation of coatings in particular, but also to the preservation of national wealth generally.

The registration process under the European Biocidal Products Regulation is complex, time-consuming, cost-intensive and harbours the risk that, despite enormous outlay, significant constraints might be imposed on biocidal actives/products or that some actives might even disappear due to a substantial administrative burden. However, a reasonably comprehensive set of actives and biocidal products based thereon is essential for covering the various preservation needs of real life and, what is more, to effectively avoid the formation of resistance by germs. Without an appropriate 'toolbox' of biocidal products featuring different modes of action, it is hard to imagine how this problem might otherwise be solved.

In the foregoing sections, an overview was provided of the different aspects of coatings protection, covering the spectrum from basic information in the universe of microorganisms, to the innate properties of microbicides, to the current state of the art and finally to legislative aspects. The intention of the author has been to provide for the reader a basic set of information to facilitate judicious choices to be made from the range of microbicidal products currently on the market. Nearly every option has pluses and negatives in this regard. Excellent performance properties, for instance, might be counterbalanced by an unfavourable classification and labelling situation and vice versa. The decision to opt for a certain microbicidal product should be broadly based and take as many aspects as possible into account. Combining actives seems to be an appropriate strategy for eliciting the maximum performance as a result of synergistic effects. A sustainable solution for a preservation problem depends on various factors and these must therefore be given due consideration. Hygienic measures during production must be considered as well.

All in all, a microbicidal product should be employed according to the basic principle:

"As much as necessary and as little as possible".

Figure 6.1: Façade deterioration (left) versus façade protection (right) – balancing all aspects in the field of preservation is essential

7 References

[1] Brill, H. (Ed.): Mikrobielle Materialzerstörung und Materialschutz. JenaStuttgart: Gustav Fischer Verlag (1995).

[2] B. V. Hofkin (2011). "Living in a Microbial World", published by Garland Science, Taylor and Francis Group, LLC, an informa business, 270 Madison Avenue, New York, NY 10016, USA, and 2 Park Square, Milton Park, Abingdon, OX14 4RN, UK, p. 233–238.

[3] E. V. Piletska*, G. Stavroulakis, K. Karim, M. J. Whitcombe, I. Chianella, A. Sharma, K. E. Eboigbodin, G. K. Robinson, S. A. Piletsky (2010). "Attenuation of Vibrio fischeri Quorum Sensing Using Rationally Designed Polymers", Biomacromolecules 11, No. 4, p. 975–980.

[4] C. H. Posten, C. L. Cooney (2001). "Growth of Microorganisms", in: Biotechnology Set, Second Edition (Eds. H.-J. Rehm and G. Reed), Wiley-VCH Verlag GmbH, Weinheim, Germany.

[5] G. W. Yeates, S. S. Bamforth, D. J. Ross, K. R. Tate, G. P. Sparling (1991). "Recolonisation of methyl bromide sterilised soils under four different field conditions", Biology and Fertility of Soils, 11(3):181–189.

[6] P. Wrede, S. Wrede (Ed.) (2012). "Charles Darwin: Die Entstehung der Arten", Kommentierte und illustrierte Ausgabe, Wiley-VCH, Weinheim, Germany.

[7] J. M. Berg, J. L. Tymoczko, L. Stryer (2002). "Biochemistry, 5th edition"; New York: W. H. Freeman; 2002. Section 2.4, Cells Can Respond to Changes in Their Environments. Available from: http://www.ncbi.nlm.nih.gov/books/NBK22568/

[8] G. W. Tannock (1995). "Normal Microflora – An Introduction to Microbes Inhabiting the Human Body", published by Chapman & Hall, 2–6 Boundary Row, London SE1 8HN, UK.

[9] Th. Warscheid (2000). "Mikrobieller Befall und Schädigung von Baustoffen", in: Biozide in Bautenbeschichtungen (E. Bagda, Ed.), Expert-Verlag, p. 10–27.

[10] R. Oswald, S. Paland, R. Spilker, K. Wilmes, J. Ihlenfeldt, A. Rietz, S. Schlösser, E. Hinz, U. Wullkopf (2003). "Systematische Instandsetzung und Modernisierung im Wohnungsbau – Endbericht", Fraunhofer IRB Verlag.

[11] Th. Warscheid, (1996). "Impacts of microbial biofilms in deteriorating inorganic building materials and their relevance for the conservation practice", Internationale Zeitschrift für Bauinstandsetzen 2(6): p. 493–504.

[12] M. Madigan, J. Martinko (Eds.) (2006). "Brock Biology of Microorganisms, 13th edition". Pearson Education, p. 1096, ISBN 0-321-73551-X.

[13] http://en.wikipedia.org/wiki/Microorganism

[14] C. R. Woese, G. E. Fox (1977). "Phylogenetic structure of the prokaryotic domain: the primary kingdoms", in: Proceedings of the National Academy of Sciences of the United States of America. Vol. 74, No. 11, p. 5088–5090.

[15] B. J. Tindall, P. Kämpfer, J. P. Euzéby, A. Oren (2006). "Valid publication of names of prokaryotes according to the rules of nomenclature: past history and current practice", in: International Journal of Systematic and Evolutionary Microbiology. Vol. 56, No. 11, p. 2715–2720, ISSN 1466-5026.

[16] J. P. Euzéby(1997). "List of bacterial names with standing in nomenclature: a folder available on the Internet", in: International Journal of Systematic Bacteriology. Vol. 47, p. 590–592. PMID 9103655.

[17] S. P. Lapage, P. H. A. Sneath, E. F. Lessel, V. B. D. Skerman, H. P. R. Seeliger, W. A. Clark (Ed.) (1992). "International Code of Nomenclature of Bacteria – Bacteriological Code", 1990 Revision. ASM Press, Washington (DC), USA, ISBN 1-55581-039-X.

[18] C. R. Woese, E. Stackebrandt, T. J. Macke, G. E. Fox (1985). "A phylogenetic definition of the major eubacterial taxa", in: Systematic and Applied Microbiology. Vol. 6, p. 143–151, ISSN 0723-2020.

[19] P. Mattarelli, W. Holzapfel (2014). "Recommended minimal standards for description of new taxa of the genera Bifidobacterium, Lactobacillus and related genera" in: International Journal of Systematic and Evolutionary Microbiology. Vol. 64, No. 4, p. 1434–1451, ISSN 1466-5026.

[20] Bergey's Manual Trust, Department of Microbiology, 527 Biological, Sciences Building, University of Georgia, Athens, GA 30602-2605, USA.

[21] W. Ludwig & 31 other authors (2004). "ARB: a software environment for sequence data", in: Nucleic acids research. Vol. 32, No. 4, p. 1363–1371, ISSN 1362-4962.

Frank Sauer: Microbicides in Coatings

[22] E. Pruesse, C. Quast & others (2007). "SILVA: a comprehensive online resource for quality checked and aligned ribosomal RNA sequence data compatible with ARB", in: Nucleic acids research. Vol. 35, No. 21, p. 7188–7196, ISSN 1362-4962.

[23] P. Yarza, M. Richter, J. Peplies, J. Euzéby, R. Amann, K. H. Schleifer, W. Ludwig, F. O. Glöckner, R. Rosselló-Móra (2008). "The All-Species Living Tree project: a 16S rRNA-based phylogenetic tree of all sequenced type strains", in: Systematic and Applied Microbiology. Vol. 31, No. 4, p. 241–250, ISSN 0723-2020.

[24] P. de Vos, H. G. Trüper (2000). Judicial Commission of the International Committee on Systematic Bacteriology. IXth International (IUMS) Congress of Bacteriology and Applied Microbiology. Minutes of the meetings, 14, 15 and 18 August 1999, Sydney, Australia. Int J Syst Evol Microbiol. 50: p. 2239–2244.

[25] C. R. Woese, O. Kandler, M. L. Wheelis (1990). "Towards a natural system of organisms: proposal for the domains Archaea, Bacteria, and Eucarya". Proceedings of the National Academy of Sciences 87 (12): p. 4576–4579.

[26] http://www.ncsu.edu/project/bio183de/Black/prokaryote/prokaryote1.html (NC State University. "Prokaryotes: Single-celled Organisms").

[27] N. Campbell "Biology: Concepts & Connections". Pearson Education. San Francisco (2003).

[28] http://www.etymonline.com/index.php?term=prokaryote&allowed_in_frame=0 ("prokaryote" – Online Etymology Dictionary).

[29] http://www.etymonline.com/index.php?term=eukaryotic&allowed_in_frame=0" ("eukaryotic" – Online Etymology Dictionary).

[30] N. R. Pace (2006), Nature, Volume 441, Issue 7091, pp. 289.

[31] R. Dunn "After 2 Years Scientists Still Can't Solve Belly Button Mystery, Continue Navel-Gazing". Guest Blog. Scientific American. Retrieved 16 January 2013.

[32] https://en.wikipedia.org/wiki/Archaea

[33] E. F. DeLong (1998). "Everything in moderation: archaea as 'non-extremophiles'". Current Opinion in Genetics & Development 8 (6): p. 649–54.

[34] J. Theron, T. E. Cloete (2000). "Molecular techniques for determining microbial diversity and community structure in natural environments". Crit. Rev. Microbiol. 26 (1): p. 37–57.

[35] T. M. Schmidt (2006). "The maturing of microbial ecology", Int. Microbiol. 9 (3): p. 217–223.

[36] K. O. Stetter (1996). "Hyperthermophiles in the history of life", Ciba Found Symp. 202: p. 1–10.

[37] L. J. Magrum, K. R. Luehrsen, C. R. Woese (1978). "Are extreme halophiles actually "bacteria"?", J Mol Evol. 11 (1): p. 1–10.

[38] http://www.nsf.gov/od/lpa/news/03/pr0384.htm (Strain 121 – Microbes from depths takes life to hottest known limit, National Science Foundation, Office of Legislative and Public Affairs, 4201 Wilson Boulevard, Arlington, Virginia 22230, US)

[39] Th. D. Brock (1997). "The Value of Basic Research: Discovery of thermus aquaticus and other extreme thermophiles", Genetics 146, p. 1207–1210.

[40] J. Seckbach, et al.: "Polyextremophiles – life under multiple forms of stress", Springer, Dordrecht (2013), ISBN 978-94-007-6487-3 available online: http://www.genetics.org/content/146/4/1207.full.pdf.

[41] https://en.wikipedia.org/wiki/Hyperthermophilic.

[42] https://en.wikipedia.org/wiki/Halophile.

[43] J. W. Schopf (1994). "Disparate rates, differing fates: tempo and mode of evolution changed from the Precambrian to the Phanerozoic". Proceedings of the National Academy of Sciences of the United States of America 91 (15): p. 6735–6742.

[44] E. F. DeLong, N. R. Pace (2001). "Environmental diversity of bacteria and archaea", Syst Biol 50 (4): p. 470–478.

[45] J. K. Fredrickson, J. M. Zachara, D. L. Balkwill, D. Kennedy, S. M. Li, H. M. Kostandarithes, M. J. Daly, M. F. Romine, F. J. Brockman (2004). "Geomicrobiology of high-level nuclear waste-contaminated vadose sediments at the Hanford site, Washington state", Applied and Environmental Microbiology 70 (7): p. 4230–4241.

[46] http://science.nasa.gov/science-news/science-at-nasa/2007/11may_locad3/

[47] W. B. Whitman, D. C. Coleman, W. J. Wiebe (1998). "Prokaryotes: the unseen majority". Proceedings of the National Academy of Sciences of the United States of America 95 (12): p. 6578–6583.

[48] C. M. Hogan (2010). "Bacteria. Encyclopedia of Earth" (Eds. S. Draggan and C.J. Cleveland), National Council for Science and the Environment, Washington DC.

[49] http://en.wikipedia.org/wiki/Bacteria

[50] Apotheken-Rundschau, (September 2015), Germany.

[51] C. Q. Choi, (2013). "Microbes Thrive in Deepest Spot on Earth". LiveScience. Retrieved 17 March 2013.

[52] R. Glud, F. Wenzhöfer, M. Middelboe, K. Oguri, R. Turnewitsch, D. E. Canfield,

H. Kitazato (2013). "High rates of microbial carbon turnover in sediments in the deepest oceanic trench on Earth". Nature Geoscience 6 (4): p. 284–288.

[53] C. L. Sears (2005). "A dynamic partnership: celebrating our gut flora". Anaerobe 11 (5): p. 247–251.

[54] M. S. Rappé, S. J. Giovannoni (2003). "The uncultured microbial majority". Annual Review of Microbiology 57: p. 369–394.

[55] M. T. Cabeen, C. Jacobs-Wagner (2005). "Bacterial cell shape". Nature Reviews Microbiology 3 (8): p. 601–610.

[56] K. D. Young (2006). "The selective value of bacterial shape". Microbiol. Mol. Biol. Rev. 70 (3): p. 660–703.

[57] https://de.wikipedia.org/wiki/Biofilm

[58] L. Hall-Stoodley, J. W. Costerton (2004). "Bacterial biofilms: from the natural environment to infectious diseases". Nature Reviews Microbiology. Bd. 2, No. 2 p. 95–108.

[59] J. W. Costerton (2007). "The Biofilm Primer" (Springer Series on Biofilms). Springer-Verlag, Berlin/Heidelberg/New York, ISBN 978-3-540-68021-5.

[60] H.-C. Flemming (2000). "Biofilme – das Leben am Rande der Wasserphase". Nachrichten aus der Chemie. 4, p. 442–447.

[61] W. C. Fuqua, S. C. Winans, E. P. Greenberg (1994). "Quorum sensing in bacteria: the LuxR-LuxI family of cell density-responsive transcriptional regulators", Journal of bacteriology. Vol. 176, No. 2.

[62] C. M. Waters, B. L. Bassler, (2005). "Quorum sensing: cell-to-cell communication in bacteria", Annu. Rev. Cell. Dev. Biol. Vol. 21, p. 319–346.

[63] U. Szewzyk, R. Szewzyk (2003). "Biofilme – die etwas andere Lebensweise", BIOspektrum. Jg. 9, p. 253–255. ISSN 0947-0867.

[64] C. M. Ott, R. J. Bruce et al. (2004). "Microbial characterisation of free floating condensate aboard the Mir space station". Microbial ecology. Vol. 47, No. 2, p. 133–136.

[65] C. Messal (2008). "Biofilme-Biofouling-Biokorrosion", MICOR Groß Lüsewitz, Internet: www.haw-hamburg.de/fileadmin/user_upload/Gleichstellung/downloads/hawscherer_vorlesung_biotech_messal_web.pdf.

[66] J. van Heijenoort (2001). "Formation of the glycan chains in the synthesis of bacterial peptidoglycan". Glycobiology 11 (3): 25R–36R.

[67] A. L. Koch (2003). "Bacterial wall as target for attack: past, present, and future research". Clin. Microbiol. Rev. 16 (4): p. 673–687.

[68] H. C. Gram (1884). "Über die isolierte Färbung der Schizomyceten in Schnitt- und Trockenpräparaten". Fortschr. Med. 2: p. 185–189.

[69] https://de.wikipedia.org/wiki/Gram-F%C3%A4rbung; (pic. by Y_tambe-Y_Tambe´s file, Attribution-ShareAlike unported (CC BY-SA 3.0))

[70] J. A. Morello, P. A. Granato, M. E. Wilson, V. Morton (2006). "Laboratory Manual and Workbook in Microbiology: Applications to Patient Care", 10th Edition Boston: McGraw-Hill Higher Education, Print.

[71] P. Varughese, D. M. Helbecque, K. B. McRae, L. Eidus (1974). "Comparison of strip and Ziehl-Neelsen methods for staining acid-fast bacteria", Bulletin of the World Health Organisation. Vol. 51, No. 1, p. 83–91.

[72] F. M. Walsh, S. G. Amyes (2004). "Microbiology and drug resistance mechanisms of fully resistant pathogens". Current Opinion in Microbiology 7(5): p. 439–444.

[73] W. L. Nicholson, N. Munakata, G. Horneck, H. J. Melosh, P. Setlow (2000). "Resistance of Bacillus endospores to extreme terrestrial and extraterrestrial environments". Microbiology and Molecular Biology Reviews 64(3): p. 548–572.

[74] W. L. Nicholson, A. C. Schuerger, P. Setlow (2005). "The solar UV environment and bacterial spore UV resistance: considerations for Earth-to-Mars transport by natural processes and human spaceflight". Mutat. Res. 571 (1–2): p. 249–264.

[75] R. H. Vreeland, W. D. Rosenzweig, D. W. Powers (2000). "Isolation of a 250 million-year-old halotolerant bacterium from a primary salt crystal". Nature 407 (6806): p. 897–900.

[76] R. J. Cano, M. K. Borucki (1995). "Revival and identification of bacterial spores in 25- to 40-million-year-old Dominican amber" Science 268 (5213): p. 1060–1064.

[77] R. M. Youngson (2006). Collins Dictionary of Human Biology. Glasgow: HarperCollins, ISBN 0-00-722134-7.

[78] D. L. Nelson, M. M. Cox (2005). Lehninger Principles of Biochemistry (4th Edition). New York: W. H. Freeman, ISBN 0-7167-4339-6.

[79] E. A. Martin (Ed.) (1983). Macmillan Dictionary of Life Sciences (2nd Edition). London: Macmillan Press, ISBN 0-333-34867-2.

[80] K. Henze, W. Martin (2003). "Evolutionary biology: essence of mitochondria". Nature 426 (6963): p. 127–128.

[81] M. Pavelk, A. A. Mironov (2008). "The Golgi Apparatus: State of the art 110 years after Camillo Golgi's discovery". Berlin: Springer, ISBN 3-211-76309-0.

[82] http://www.etymonline.com/index.php?term=chloroplast&allowed_in_frame=0 "chloroplast". Online Etymology Dictionary.

[83] http://en.wikipedia.org/wiki/Chloroplast

[84] F. Huber, J. Schnauss, S. Roenicke, P. Rauch, K. Mueller, C. Fuetterer, J. Kaes (2013) "Emergent complexity of the cytoskeleton: from single filaments to tissue". Advances in Physics. 62, No. 1, p. 1–112.

[85] J. Lomako, W. M. Lomako, W. J. Whelan (2004). "Glycogenin: the primer for mammalian and yeast glycogen synthesis", Biochim. Biophys. Acta. Vol. 1673, p. 45–55.

[86] S. M. Bowman, S. J. Free (2006). "The structure and synthesis of the fungal cell wall", Bioessays. Vol. 28, p. 799–808.

[87] P. Sitte, H. Ziegler, F. Ehrendorfer (1991). "Strasburger Lehrbuch der Botanik", 33rd Edition Urban & Fischer.

[88] https://de.wikipedia.org/wiki/Pilze

[89] J. M. Barea, M. J. Pozo, R. Azcón, C. Azcón-Aguilar (2005). "Microbial co-operation in the rhizosphere", J. Exp. Bot. Vol. 56, p. 1761–1778.

[90] P. Sitte, E. Weiler, J. W. Kadereit, A. Bresinsky, C. Körner (2002). "Lehrbuch der Botanik für Hochschulen". Begründet von Eduard Strasburger. 35th Edition Spektrum Akademischer Verlag, Heidelberg, p. 353–356.

[91] U. Paszkowski (2006). "Mutualism and parasitism: the yin and yang of plant symbioses", Curr. Opin. Plant Biol. Vol. 9, p. 364–370.

[92] M. Paoletti, K. W. Buck, C. M. Brasier (2006). "Selective acquisition of novel mating type and vegetative incompatibility genes via interspecies gene transfer in the globally invading eukaryote Ophiostoma novo-ulmi", Mol Ecol. Vol. 15, p. 249–262.

[93] C. J. Alexopoulos, C. W. Mims, M. Blackwell (1996). "Introductory Mycology". John Wiley and Sons.

[94] S. D. Harris (2008). "Branching of fungal hyphae: regulation, mechanisms and comparison with other branching systems". Mycologia 100 (6): p. 823–832.

[95] C. L. Schmitt, M. L. Tatum (2008). "The Malheur National Forest. Location of the World's Largest Living Organism". The Humongous Fungus. United States Department of Agriculture.

[96] M. Blackwell (2011). "The Fungi: 1, 2, 3 ... 5.1 million species?", American Journal of Botany. 98, No. 3, p. 426–438.

[97] E. Dadachova, R. A. Bryan, X. Huang, T. Moadel, A. D. Schweitzer, P. Aisen, J. D. Nosanchuk, A. Casadevall (2007). "Ionizing Radiation Changes the Electronic Properties of Melanin and Enhances the Growth of Melanized Fungi", PloS ONE 2(5), p. 457.

[98] https://de.wikipedia.org/wiki/Melanine

[99] http://www.biomess.eu/mvoc.htm

[100] https://en.wikipedia.org/wiki/Geosmin

[101] Landesgesundheitsamt Baden-Württemberg (2004). "Schimmelpilze in Innenräumen – Nachweis, Bewertung, Qualitätsmanagement", Abgestimmtes Arbeitsergebnis des Arbeitskreises "Qualitätssicherung – Schimmelpilze in Innenräumen" am Landesgesundheitsamt Baden-Württemberg 14.12.2001 (überarbeitet Dezember 2004).

[102] https://de.wikipedia.org/wiki/Alge

[103] C. v. d. Hoek, D. G. Mann, H. M. Jahns (1995). "Algae: an introduction to phycology", Cambridge University Press (623 pp).

[104] M. Allaby, (Ed.) (1992). "Algae". The Concise Dictionary of Botany. Oxford: Oxford University Press.

[105] M. W. Nabors (2004). "Introduction to Botany", San Francisco, CA: Pearson Education, Inc., ISBN 978-0-8053-4416-5.

[106] P. C. Silva, R. L. Moe (2014). "Cyanophyceae", AccessScience. McGraw-Hill Education, (Web. 10 Nov. 2015).

[107] R. E. Lee (2008). "Phycology", Cambridge University Press.

[108] http://www.umweltdatenbank.de/cms/lexikon/27-lexikon-a/587-algen.html

[109] http://www.spektrum.de/lexikon/biologie/algen/2029

[110] http://www.nationalgeographic.de/aktuelles/algen-kommen-ohne-licht-und-sauerstoff-aus

[111] C. A. Crispim, P. M. Gaylarde, Ch. C. Gaylarde (2003). "Algal and Cyanobacterial Biofilms on Calcareous Historic Buildings", Current Microbiology Vol. 46, p. 79–82.

[112] C. Grant (1982). "Fouling of terrestrial substrates by algae and implications for control – a review", Int. Biodeterior Bull 18, p. 57–65.

[113] J. Braun-Blanquet (1951). "Pflanzensoziologie – Grundzüge der Vegetationskunde", Springer-Verlag Wien GmbH, p. 114–115, ISBN 978-3-7091-4079-6.

[114] H. Brauer (Ed.) (1996). "Handbuch des Umweltschutzes und der Umweltschutztechnik – Vol. 1: Emissionen und ihre Wirkungen", Springer-Verlag Berlin Heidelberg, p. 456, ISBN 978-3-642-63892-3.

[115] K. Esser (2000). "Kryptogamen 1 – Cyanobakterien, Algen, Pilze, Flechten", Springer-Verlag Berlin Heidelberg, p. 61, ISBN 978-3-642-63056-9.

[116] G. Caneva, M. P. Nugari, O. Salvadori (2008). "Plant Biology for Cultural Heritage – Biodeterioration and Conservation", The Getty Conversation Institute, Getty Publications, 1200 Getty Centre Drive, Suite 500, Los Angeles, California 90049-1682.

[117] F. Gladis, R. Schumann, "Wirksamkeit von photokatalytischen Oberflächen gegen Algenbewuchs auf Fassaden" (Abschlussbericht

zum Max-Buchner-Forschungsprojekt), Universität Rostock, Institut für Biowissenschaften, Angewandte Ökologie, Albert-Einstein-Str. 3, 18051 Rostock; available online: http://www.dechema.de/dechema_media/2786.pdf.

[118] F. Gladis, A. Eggert, U. Karsten, R. Schumann (2010). "Prevention of biofilm growth on man-made surfaces – Evaluation of antialgal activity of two biocides and photocatalytic nanoparticles", Biofouling, 26, p. 89–101.

[119] S. Scheerer, O. Ortega-Morales, C. Gaylarde (2009). "Microbial Deterioration of Stone Monuments – An Updated Overview", Advances in Applied Microbiology 66, p. 97–139.

[120] J. I. S. Khattar, D. P. Singh, G. Kaur, (2009). "Algal Biology and Biotechnology", I. K. International Publishing House Pvt. Ltd., S-25, Green Park Extension, Uphaar Cinema Market, New Delhi, Bangalore (India), ISBN 978-93-80026-19-0.

[121] J. Seckbach (2007). "Algae and Cyanobacteria in Extreme Environments", published by Springer, P. O. Box 17, 3300 AA Dordrecht, The Netherlands, p. 632, ISBN 978-1-4020-6111-0 (HB).

[122] K. Mörbe, W. Morenz, H.-W. Pohlmann, H. Werner (1987). "Praktischer Korrosionsschutz: Korrosionsschutz wasserführender Anlagen", erweiterter Nachdruck der zweiten bearbeiteten Auflage, Springer-Verlag Wien – New York.

[123] H. Ettl, G. Gärtner (2014). "Syllabus der Boden-, Luft- und Flechtenalgen", 2nd Edition, Gustav Fischer Verlag, Stuttgart, p. 773.

[124] https://de.wikipedia.org/wiki/Flechte

[125] http://medical-dictionary.thefreedictionary.com/microbicide

[126] Regulation (EU) No. 528/2012 of the European Parliament and of the Council of 22 May 2012 concerning the making available on the market and use of biocidal products (consolidated version dated 25.04.2014); available online: http://eur-lex.europa.eu/legal-content/EN/TXT/PDF/?uri=CELEX:02012R0528-20140425&rid=2.

[127] W. Paulus (2005). "Directory of Microbicides for the Protection of Materials – A Handbook", published by Springer, P.O. Box 17, 3300 AA Dordrecht, The Netherlands.

[128] F. Myers (2008). "Biocidal agents: modes of action and correlation with antibiotic resistance", The Biomedical Scientist, pp. 227–231.

[129] Fungicide Resistance Action Committee (FRAC), FRAC Code List © 2015. "Fungicides sorted by mode of action (including FRAC Code numbering)", available online: http://www.frac.info/docs/default-source/publications/frac-code-list/frac-code-list-2015-finalC2AD7AA36764.pdf.

[130] A. P. Fraise, J.-Y. Maillard, S. A. Sattar (Eds.) (2013). "Russell, Hugo and Ayliffe's Principles and Practice of Disinfection, Preservation and Sterilisation, 5th edition, Wiley-Blackwell, ISBN 1-4443-3325-9.

[131] Scientific Committee on Emerging and Newly Identified Health Risks (SCENIHR) (2009). "Assessment of the Antibiotic Resistance Effects of Biocides" (SCENIHR adopted this opinion at the 28th plenary on 19 January 2009 after public consultation), available online: http://ec.europa.eu/health/scientific_committees/emerging/opinions/scenihr_opinions_en.htm.

[132] A. J. Carrillo-Muñoz, G. Giusiano, P. A. Ezkurra, G. Quindós (2006). "Antifungal agents: Mode of action in yeast cells", Rev Esp Quimioterap, 19(2): p. 130–139.

[133] J. Stenersen (2004). "Chemical Pesticides Mode of Action and Toxicology", CRC Press Boca Raton, London, New York, Washington D. C., ISBN 0-7484-0910-6.

[134] J.-Y. Maillard (2002). "Bacterial target sites for biocide action", Journal of Applied Microbiology Symposium Supplement 92, 16S–27S.

[135] M. A. Ghannoum, L. B. Rice (1999). "Antifungal Agents: Mode of Action, Mechanisms of Resistance, and Correlation of These Mechanisms with Bacterial Resistance", Clinical Microbiology Reviews, 12(4): p. 501–517.

[136] S. P. Denyer, G. S. A. B. Stewart (1998). "Mechanisms of Action of Disinfectants", International Biodeterioration & Biodegradation, 41: p. 261–268.

[137] A. D. Russell, J. R. Furr, J.-Y. Maillard (1997). "Microbial susceptibility and resistance to biocides". ASM News, 63: p. 481–487.

[138] W. B. Hugo (1995). "A brief history of heat, chemical and radiation preservation and disinfection", International Biodeterioration & Biodegradation, 36(3): p. 197–217.

[139] S. P. Denyer (1995). "Mechanisms of Action of Antibacterial Biocides", International Biodeterioration & Biodegradation, 36: p. 227–245.

[140] S. P. Denyer, W. B. Hugo (Eds.) (1991). "Mechanisms of Action of Chemical Biocides: Their Study and Exploitation", Blackwell Scientific Publications, Oxford, United Kingdom, ISBN 0-632-02928-5.

[141] Centres for Disease Control and Prevention, 1600 Clifton Rd, Atlanta, GA 30333, Guideline for Disinfection and Sterilisation in Healthcare Facilities, 2008, available online:

http://www.cdc.gov/hicpac/Disinfection_Sterilisation/6_0disinfection.html.
[142] K. Park Talaro (2007). "Foundations in Microbiology – Basic Principles", Sixth Edition, published by McGraw-Hill, New York.
[143] G. McDonnell, A. D. Russell (1999). "Antiseptics and Disinfectants: Activity, Action, and Resistance". Clinical Microbiology Reviews, 12(1): p. 147–179.
[144] P. M. Davidson (1997). "Chemical preservatives and natural antimicrobial compounds", In Food Microbiology Fundamentals and Frontiers, (Eds. M. P. Doyle, L. R. Beuchat, T. J. Montville), ASM Press, New York, pp. 520–556.
[145] S. E. Walsh, J.-Y. Maillard, A. D. Russell, C. E. Catrenich, D. L. Charbonneau, R. G. Bartolo (2003). "Activity and mechanisms of action of selected biocidal agents on Gram-positive and -negative bacteria", Journal of Applied Microbiology, 94: p. 240–247.
[146] N. C. Rath, K. S. Rasaputra, R. Liyanage, G. R. Huff, W. E. Huff (2011). "Dithiocarbamate Toxicity – An Appraisal, Pesticides in the Modern World – Effects of Pesticides Exposure", (Ed. Dr. Margarita Stoytcheva), InTech, ISBN 978-953-307-454-2, available from: http://www.intechopen.com/books/pesticides-in-the-modern-world-effects-of-pesticidesexposure/dithiocarbamate-toxicity-an-appraisal.
[147] F. Leonard, F. A. Barkley, E. V. Brown, F. E. Anderson, D. M. Green (1956). "The relationship of structure to microbiologic activity in the N-oxides". Antibiotics Chemotherapy, 6: p. 261.
[148] S. M. Abdel-Malek, I. S. Al-Adham, C. L. Winder, T. E. Buultjens, K. M. Gartland, P. J. Collier (2002). "Antimicrobial susceptibility changes and T-OMP shifts in pyrithione-passaged planktonic cultures of Pseudomonas aeruginosa PAO1", Journal of Applied Microbiology 92: p. 729–736.
[149] J. J. Cooney, J. A. Felix (1972). "Inhibition of Cladosporium resinae in hydrocarbon-water systems by pyridinethiones", Int. Biodet. Bull., 8(2): p. 59–63.
[150] C. S. Chandler, I. H. Segel (1978). "Mechanism of the antimicrobial action of pyrithione: effects on membrane transport, ATP levels and protein synthesis", Antimicrob. Agents Chemother. 14(1): p. 60–68.
[151] T. Kato (1986). "Sterol biosynthesis in fungi, a target for broad spectrum fungicides", Chemistry of Plant Protection (Germany, F.R.), available online: http://agris.fao.org/agris-search/search.do?recordID=DE87U0023.
[152] I. M.-Y. Kwok, R. T. Loeffler (1993). "The biochemical mode of action of some newer azole fungicides", Pestic. Sci, 39: p. 1–11.
[153] S. L. Kelly, D. C. Lamb, A. J. Corran, B. C. Baldwin, D. E. Kelly (1995). "Mode of Action and Resistance to Azole Antifungals Associated with the Formation of 14α-Methylergosta-8,24(28)-dien-3β,6α-diol", Biochemical and Biophysical Research Communications, 207(3): p. 910–915.
[154] E. Lacey (1990). "Mode of Action of Benzimidazoles", Parasitology Today, 6(4): p. 112–115.
[155] L. C. Davidse (1986). "Benzimidazole Fungicides: Mechanism of Action and Biological Impact", Annual Review of Phytopathology, 24: p. 43–65.
[156] L. C. Davidse (1977). "Mode of Action, Selectivity and Mutagenicity of Benzimidazole Compounds", Netherlands Journal of Plant Pathology, 83 (Suppl. 1): p. 135–144.
[157] S. Gomathinayagam, N. Balasubramanian, V. Shanmugaiah, M. Rekha, P. T. Manoharan, D. Lalithakumari (2011). "Molecular Characterisation of Carbendazim Resistance of Plant Pathogen (Bipolaris oryzae)", Fungicides – Beneficial and Harmful Aspects, (Ed. Dr. Nooruddin Thajuddin), InTech, ISBN 978-953-307-451-1, available from: http://www.intechopen.com/books/fungicides-beneficial-and-harmfulaspects/molecular-characterisation-of-carbendazim-resistance-of-plant-pathogen-bipolaris-oryzae.
[158] P. J. Collier, A. Ramsey, R. D. Waigh, K. T. Douglas, P. Austin, P. Gilbert (1990). "Chemical reactivity of some isothiazolone biocides", J. Appl. Bacteriol. 69: p. 578–584.
[159] T. M. Williams (2007). "The Mechanism of Action of Isothiazolone Biocides", PowerPlant Chemistry 9(1): p. 14–22.
[160] R. J. Stretton, T. W. Manson (1973). "Some Aspects of the Mode of Action of the Antibacterial Compound Bronopol (2-bromo-2-nitropropan-1,3-diol)", Journal of Applied Bacteriology, 36: p. 61–76.
[161] J. A. Shepherd, R. D. Waigh, P. Gilbert (1988). "Antibacterial action of 2-bromo-2-nitropropane-1,3-diol (bronopol)", Antimicrobial Agents and Chemotherapy, 32(11): p. 1693–1698.
[162] P. Gilbert, L. E. Moore (2005). "A Review – Cationic antiseptics: diversity of action under a common epithet", Journal of Applied Microbiology, 99: p. 703–715.
[163] P. Maris (1995). "Modes of action of disinfectants", Rev. sci. tech. Off. int. Epiz., 14 (1): p. 47–55.
[164] M. Hostovsky, J. Blahova, L. Plhalova, V. Kopriva, Z. Svobodova (2014). "Effects of the exposure of fish to triazine herbicides", Neuroendocrinology Letters, 35(Suppl. 2): p. 3–25.

[165] A. Kositchaiyong, V. Rosarpitak, B. Prapagdee, N. Sombatsompop (2013). "Molecular characterisations, mechanical properties and anti-algal activities for PVC and wood/PVC composites containing urea- and triazine-based algaecides", Composites Part B: Engineering, 53:25–35, available online: http://www.sciencedirect.com/science/article/pii/S1359836813001844.

[166] R. Swindell (2003). "Shining light on the evolution of photosynthesis", Answers Magazine, 17: p. 74 – 84.

[167] J. G. Metz, H. B. Pakrasi, M. Seibert, C. J. Arntzen (1986). "Evidence for a dual function of the herbicide-binding D1 protein in photosystem II", FEBS Letters, 205: p. 269 – 274.

[168] K. H. Büchel (1972). "Mechanisms of action and structure activity relations of herbicides that inhibit photosynthesis", Pesticide Science, 3(1): p. 89 – 110.

[169] K. Pfister, C. J. Arntzen (1979). "The Mode of Action of Photosystem II-Specific Inhibitors in Herbicide-Resistant Weed Biotypes", Z. Naturforsch. 34(c): p. 996 – 1009.

[170] J. J. Schleier III, R. K. D. Peterson (2012). "The joint toxicity of type I, II, and non-ester pyrethroid insecticides", Journal of Economic Entomology 105: p. 85 – 91.

[171] J. J. Schleier III, R. K. D. Peterson (2011). "Pyrethrins and pyrethroid insecticides", in: O. Lopez and J. G. Fernández-Bolanos (Eds.), Green trends in insect control, Royal Society of Chemistry, London, United Kingdom, p. 94 – 131.

[172] D. M. Soderlund, J. M. Clark, L. P. Sheets, L. S. Mullin, V. J. Piccirillo, D. Sargent, J. T. Stevens, M. L. Weiner (2002). "Mechanisms of pyrethroid neurotoxicity: implications for cumulative risk assessment", Toxicology, 171(1): p. 3 – 59, available online: http://www.sciencedirect.com/science/article/pii/S0300483X01005698.

[173] H. P. M. Vijverberg, J. van den Bercken (1990). "Neurotoxicological Effects and the Mode of Action of Pyrethroid Insecticides", Critical Reviews in Toxicology, 21(2): p. 105 – 126.

[174] S. Prabhu, E. K. Poulose (2012). "Silver nanoparticles: mechanism of antimicrobial action, synthesis, medical applications, and toxicity effects", International Nano Letters, 2(1): p. 1 – 10.

[175] K.-J. Kim, W. S. Sung, B. K. Suh, S.-K. Moon, J.-S. Choi, J. G. Kim, D. G. Lee (2009). "Antifungal activity and mode of action of silver nano-particles on Candida albicans", Biometals, 22: p. 235 – 242.

[176] D. Paul, S. Paul, N. Roohpour, M. Wilks, P. Vadgama (2013). "Antimicrobial, Mechanical and Thermal Studies of Silver Particle-Loaded Polyurethane", J. Funct. Biomater. 4: p. 358 – 375.

[177] G. Zhao, S. E. Stevens Jr. (1998). "Multiple parameters for the comprehensive evaluation of the susceptibility of Escherichia coli to the silver ion", Biometals, 11(1): p. 27 – 32.

[178] A. D. Russell, W. B. Hugo (1994). "Antimicrobial activity and action of silver", Progress in Medical Chemistry 31: p. 351 – 370.

[179] R. Slawson, J. T. Trevors, H. Lee (1992). "Silver accumulation and resistance in pseudomonas stutzeri", Archives of Microbiology, 158(6): p. 398 – 404.

[180] Article 95 list published by the European Chemicals Agency (ECHA) on their website, available online: http://echa.europa.eu/information-on-chemicals/active-substance-suppliers (09.05.2017).

[181] Directive 98/8/EC of the European Parliament and of the Council of 16 February 1998 concerning the placing of biocidal products on the market (consolidated version dated 20.08.2013); available online: http://eur-lex.europa.eu/legal-content/EN/TXT/PDF/?uri=CELEX:01998L0008-20130820&rid=1.

[182] Commission Delegated Regulation (EU) No. 1062/2014 of 4 August 2014 on the work programme for the systematic examination of all existing active substances contained in biocidal products referred to in Regulation (EU) No. 528/2012 of the European Parliament and of the Council, Official Journal of the European Union L 294/1 (10. 10. 2014); available online: http://eur-lex.europa.eu/legal-content/EN/TXT/?uri=CELEX%3A32014R1062.

[183] H. Kalász (2003). "Biological role of formaldehyde, and cycles related to methylation, demethylation and formaldehyde production", Mini Rev. med. Chem., 3, p. 175 – 192.

[184] IARC Monographs on the Evaluation of Carcinogenic Risks to Humans (2006). "Formaldehyde, 2-Butoxyethanol and 1-tert-Butoxypropan-2-ol", 88: p. 48 – 49; available online: http://monographs.iarc.fr/ENG/Monographs/vol88/index.php.

[185] Regulation (EC) No. 1223/2009 of the European Parliament and of the Council of 30 November 2009 on cosmetic products (consolidated version dated 03.03.2017); available online: http://eur-lex.europa.eu/legal-content/EN/TXT/PDF/?uri=CELEX:02009R1223-20170303.

[186] C. Lv, J. Hou, W. Xie, H.Cheng (2015). "Investigation on formaldehyde release from

preservatives in cosmetics", International Journal of Cosmetic Science, 37: p. 474–478.

[187] Commission Regulation (EU) No. 605/2014 of 5 June 2014 amending, for the purposes of introducing hazard and precautionary statements in the Croatian language and its adaptation to technical and scientific progress, Regulation (EC) No. 1272/2008 of the European Parliament and of the Council on classification, labelling and packaging of substances and mixtures (consolidated version dated 25.03.2015); available online: http://eur-lex.europa.eu/legal-content/EN/TXT/?qid=1456688761842&uri=CELEX:02014R0605-20150325.

[188] Commission Regulation (EU) 2015/491 of 23 March 2015 amending Regulation (EU) No. 605/2014 amending, for the purposes of introducing hazard and precautionary statements in the Croatian language and its adaptation to technical and scientific progress, Regulation (EC) No. 1272/2008 of the European Parliament and of the Council on classification, labelling and packaging of substances and mixtures; available online: http://eur-lex.europa.eu/legal-content/EN/TXT/?qid=1456689286465&uri=CELEX:32015R0491.

[189] W. Paulus (1988). "Development in microbicides for the protection of materials", in: D. R. Houghton, R. N. Smith and H. O. W. Eggins, (Eds.) Biodeterioration 7, Elsevier Applied Science Publishers, London pp. 1–19.

[190] M. A. Diehl, J. S. Chapman (1999). "Association of the biocide 5-chloro-2-methylisothiazol-3-one with Pseudomonas aeruginosa and Pseudomonas fluorescens", Int. Biodeterioration and Biodegradation, 44: p. 191–199.

[191] T. Köhler, A. Carlsen, G. Tiedtke (2011). "Breakthrough innovation for in-can preservation", Polymer Paint Colour Journal (PPCJ), 201(4563): p. 28–29.

[192] European Commission, Directorate-General for Health and Food Safety, Scientific Committee on Consumer Safety (SCCS) (2009). "Opinion on the mixture of 5-chloro-2-methylisothiazolin-3(2H)-one and 2-methylisothiazolin-3(2H)-one" – COLIPA No. P56; available online: http://ec.europa.eu/health/scientific_committees/consumer_safety/docs/sccs_o_009.pdf.

[193] European Commission, Directorate-General for Health and Food Safety, Scientific Committee on Consumer Safety (SCCS) (2012). "Opinion on Benzisothiazolinone" – COLIPA No. P96; available online: http://ec.europa.eu/health/scientific_committees/consumer_safety/docs/sccs_o_099.pdf.

[194] http://bio.lonza.com/uploads/tx_mwaxmarketingmaterial/Lonza_Brochures_Building_Products_Preservation_Selection_Guide_Europe.pdf.

[195] http://bio.lonza.com/uploads/tx_mwaxmarketingmaterial/Lonza_Brochures_Building_Products_Preservative_Selection_Guide_Global.pdf.

[196] Regulation (EC) No. 1272/2008 of the European Parliament and of the Council on classification, labelling and packaging of substances and mixtures (consolidated version dated 01.01.2016); available online: http://eur-lex.europa.eu/legal-content/EN/TXT/PDF/?uri=CELEX:02008R1272-20160101&rid=1.

[197] CEPE Guidance on labelling decorative paint with skin sensitising biocides – Final (28.10.2014); available online: http://sveff.se/files/2015/04/CEPE-guidance-on-labelling-with-biocides-skin-sensitisers-final-28-Oct-14.pdf.

[198] T. Köhler, A. Carlsen, G. Tiedtke (2011). "New solutions for in-can preservation in the making for Europe", European Coatings Conference (ECC), Novel Biocide Technology V, Proceedings p. 65–75.

[199] Submitted CLH proposal; Dossier submitted by Poland (December 2015); available online: http://echa.europa.eu/de/registry-of-submitted-harmonised-classification-and-labelling-intentions/-/substance-rev/12105/del/50/col/synonymDynamicField_235/type/desc/pre/1/view.

[200] G. Y. Legin (1996). "2-Bromo-2-nitro-1,3-propanediol (Bronopol) and its derivatives: synthesis, properties and application (a review)", Pharmaceutical Chemistry Journal, 30(4): p. 273–284.

[201] G. Muyzer, A. J. M. Stams (2008). "The ecology and biotechnology of sulphate-reducing bacteria", Nature Reviews Microbiology, 6: p. 441–454.

[202] M. Kireche, J.-L. Peiffer, D. Antonios, I. Fabre, E. Giménez-Arnau, M. Pallardy, J.-P. Lepoittevin, J.-C. Ourlin (2011). "Evidence for Chemical and Cellular Reactivities of the Formaldehyde Releaser Bronopol, Independent of Formaldehyde Release", in: Chemical Research in Toxicology. 24(12): p. 2115–2128.

[203] K. Kajimura, T. Tagami, T. Yamamoto, S. Iwagami (2008). "The Release of Formaldehyde upon Decomposition of 2-Bromo-2-nitropropan-1,3-diol (Bronopol)", Journal of Health Science, 54(4): p. 488–492.

[204] 28th meeting of representatives of Member States Competent Authorities for the implementation of Directive 98/8/EC concerning the placing of biocidal products

on the market (CA-Feb08-Doc.8.4-rev2), "Conclusions of a workshop on formaldehyde releasers, held in Warsaw on 14–15 January 2008", available online: https://circabc.europa.eu/faces/jsp/extension/wai/dialog/container.jsp?FormPrincipal:_idcl=FormPrincipal:SearchContentList:pager&page=2&FormPrincipal_SUBMIT=1&org.apache.myfaces.trinidad.faces.STATE=DUMMY.

[205] 57th meeting of representatives of Member States Competent Authorities for the implementation of Regulation 528/2012 concerning the making available on the market and use of biocidal products (CA-Sept14-Doc.4.1-rev1), "Management of in situ generated active substances in the context of the BPR", available online: https://circabc.europa.eu/faces/jsp/extension/wai/dialog/container.jsp?FormPrincipal:_idcl=FormPrincipal:SearchContentList:pager&page=1&FormPrincipal_SUBMIT=1&org.apache.myfaces.trinidad.faces.STATE=DUMMY.

[206] A. Kramer, O. Assadian (2008). "Wallhäußers Praxis der Sterilisation, Antiseptik und Konservierung", Georg Thieme Verlag KG, Stuttgart, ISBN 978-3-131411-211.

[207] Desinfektionsmittel-Datenbank der Stadt Wien (Bewerteter Inhaltsstoff: 2-Bromo-2-nitropropan-1,3-diol), as of March 2016, available online: https://www.wien.gv.at/wuawides/internet/Inhaltsstoffsuche/Detail/290.

[208] P. Wachtler (2007). "Evaluation of formaldehyde emissions according to the emission chamber test with bronopol-equipped wall paints", Internal report Lanxess Deutschland GmbH, Germany (unpublished).

[209] F. Sauer (2007). "Analysis report for the formaldehyde determination in the gas phase of paint buckets", internal report Lanxess Deutschland GmbH, Germany (unpublished).

[210] Commission Implementing Regulation (EU) 2016/1086 of 5 July 2016 approving 2-bromo-2-(bromomethyl)pentanedinitrile (DBDCB) as an existing active substance for use in biocidal products of product-type 6, available online: http://eur-lex.europa.eu/legal-content/EN/ALL/?uri=CELEX:32016R1086.

[211] P. A. Wolf, P. W. Sterner (1972). "2,2-Dibromo-3-nitrilopropionamide, a Compound with Slimicidal Activity". Appl. Environ. Microbiol. 24(4): p. 581–584.

[212] Th. Warscheid, J. Braams (2000). "Biodeterioration of Stones – A Review", International Biodeterioration and Biodegradation 46: p. 343–368.

[213] W. Hofbauer, C. Fitz, M. Krus, K. Sedlbauer, K. Breuer (2006). "Prognoseverfahren zum biologischen Befall durch Algen, Pilze und Flechten an Bauteiloberflächen auf Basis bauphysikalischer und mikrobieller Untersuchungen", Bauforschung für die Praxis, Vol. 77, Fraunhofer IRB Verlag.

[214] M. Burkhard, R. Kägi, H. Simmler, A.Ulrioch, M. Boller, (2007). "Nanopartikel auf Fassaden erforschen", PUSCH – Thema Umwelt: Nanotechnologie: Chancen und Risiken für die Umwelt.

[215] Th. Warscheid (2002). "Sie kommen nicht von allein", Im Fokus: Algen und Pilze, Die Mappe 11, p. 8–13.

[216] Wissenschaftliche Dienste des Deutschen Bundestages: Nachhaltigkeit, Der aktuelle Begriff 06/2004, 6 April 2004.

[217] European Chemicals Agency (ECHA) Website, Information on Chemicals-Biocidal Active Substances (10.05.2017), available online: http://echa.europa.eu/information-on-chemicals/biocidal-active-substances.

[218] Assessment Report under Regulation (EU) No. 528/2012, Evaluation of the active substance 3-iodo-2-propynyl butyl carbamate (IPBC) for product-type 13 (working or cutting fluid preservatives), evaluating Competent Authority: Denmark (January 2015); available online: http://echa.europa.eu/information-on-chemicals/biocidal-active-substances.

[219] Assessment Report, Inclusion of active substances in Annex I or IA to Directive 98/8/EC, Thiabendazole (TBZ) for product-type 8 (wood preservatives), evaluating Competent Authority: Spain (22 February 2008); available online: http://echa.europa.eu/information-on-chemicals/biocidal-active-substances.

[220] EFSA (European Food Safety Authority) 2014. "Conclusion on the peer review of the pesticide risk assessment of the active substance thiabendazole", EFSA Journal 2014; 12 (11): p. 3880; available online: http://www.efsa.europa.eu/sites/default/files/scientific_output/files/main_documents/3880.pdf.

[221] EXTOXNET (Extension Toxicology Network), Pesticide Information Profiles, (primary files maintained and archived at Oregon State University), Revised June 1996, available online: http://extoxnet.orst.edu/pips/thiabend.htm.

[222] EFSA (European Food Safety Authority) 2010. "Conclusion on the peer review of the pesticide risk assessment of the active substance carbendazim", EFSA Journal 2010; 8(5): p. 1598; available online: http://www.efsa.europa.eu/sites/default/files/scientific_output/files/main_documents/1598.pdf.

[223] European Commission, Directorate D – Food Safety: Production and distribution chain, Unit D.3 – Chemicals, contaminants and pesticides (2007). "Review report for the active substance carbendazim".

[224] G. P. Clemons, H. B. Sisler (1971). "Localization of the site of action of a fungitoxic benomyl derivative", Pestic. Biochem. Biophys. 1: p. 32–43.

[225] R. S. Hammerschlag (1973). "Benomyl and methyl-2-benzimidazole carbamate (MBO): Biochemical, cytological and chemical aspects of toxicity to Ustilago maydis and Saccharomyces cerevisiae", Diss. Abstracts Int. 33(11): 5092B (Reviewed in Abstract in Health Aspect of Pesticides No. 73-2456).

[226] European Chemicals Agency (ECHA) Website, REACH registration dossier: zinc pyrithione, available online: http://echa.europa.eu/registration-dossier/-/registered-dossier/14333.

[227] Scientific Committee on Consumer Safety (SCCS), opinion on zinc pyrithione (COLIPA No. P81), revision of 18 June 2014, available online: http://ec.europa.eu/health/scientific_committees/consumer_safety/docs/sccs_o_133.pdf.

[228] European Chemicals Agency (ECHA) Website: Submitted CLH proposal for pyrithione zinc (26.10.2016), available online: https://echa.europa.eu/en/registry-of-submitted-harmonised-classification-and-labelling-intentions/-/substance-rev/15343/term.

[229] J. J. Florio, D. J. Miller (Eds.) (2004) "Handbook of Coatings Additives" (2nd edition), Marcel Dekker, Inc, New York – Basel, p. 280.

[230] United States Environmental Protection Agency (EPA), Reregistration Eligibility Decision for 2-Octyl-3 (2H)-isothiazolone (OIT), September 2007, available online: https://archive.epa.gov/pesticides/reregistration/web/pdf/octhilinone-red.pdf.

[231] National Pesticide Information Retrieval System, available online: http://npirspublic.ceris.purdue.edu/ppis/.

[232] European Chemicals Agency (ECHA) Website: Biocidal Products Committee opinions on active substance approval, adopted BPC opinion on OIT for PT8 (15.12.2016); available online: https://echa.europa.eu/regulations/biocidal-products-regulation/approval-of-active-substances/bpc-opinions-on-active-substance-approval?p_p_id=viewsubstances.

[233] Assessment Report under Regulation (EU) No. 528/2012, Evaluation of the active substance 4,5-Dichloro-2-octyl-2H-isothiazol-3-one (DCOIT) for product-type 21 (antifouling products), evaluating Competent Authority: Norway (March 2014); available online: http://echa.europa.eu/information-on-chemicals/biocidal-active-substances.

[234] Assessment Report, Inclusion of active substances in Annex I to Directive 98/8/EC, 4,5-Dichloro-2-octyl-2H-isothiazol-3-one (DCOIT) for product-type 8 (wood preservatives), evaluating Competent Authority: Norway (February 2011); available online: http://echa.europa.eu/information-on-chemicals/biocidal-active-substances.

[235] Submitted CLH proposal: DCOIT; Dossier submitted by Norway (February 2016); available online: http://echa.europa.eu/registry-current-classification-and-labelling-intentions/-/substance-rev/12435/term.

[236] ctgb, BIJLAGE II bij het besluit d.d. 24 mei 2013 tot toelating van het middel Densil DN, toelatingnummer 14082 N; available online: http://www.ctb.agro.nl/ctb_files/14082_02.html.

[237] EFSA (European Food Safety Authority) 2014. "Conclusion on the peer review of the pesticide risk assessment of the active substance tebuconazole", EFSA Journal 2014;12(1): p. 3485; available online: http://www.efsa.europa.eu/de/efsajournal/pub/3485.

[238] Assessment Report under Regulation (EU) No.528/2012, Evaluation of the active substance Tebuconazole for product-type 7 (film preservatives), evaluating Competent Authority: Denmark (September 2013); available online: http://echa.europa.eu/information-on-chemicals/biocidal-active-substances.

[239] Commission Implementing Regulation (EU) No. 1038/2013 of 24 October 2013. "Approving tebuconazole as an existing active substance for use in biocidal products for product-types 7 and 10", Official Journal of the European Union L 283/40 (25.10.2013); available online: http://eur-lex.europa.eu/LexUriServ/LexUriServ.do?uri=OJ:L:2013:283:0040:0042:EN:PDF.

[240] Assessment Report, Inclusion of active substances in Annex I to Directive 98/8/EC, Tebuconazole for product-type 8 (wood preservatives), evaluating Competent Authority: Denmark (29 November 2007); available online: http://echa.europa.eu/information-on-chemicals/biocidal-active-substances.

[241] Assessment Report under Regulation (EU) No. 528/2012, Evaluation of the active substance Tebuconazole for product-type 10 (construction material preservatives), evaluating Competent Authority: Denmark (September 2013); available online:

http://echa.europa.eu/information-on-chemicals/biocidal-active-substances.
[242] Assessment Report under Regulation (EU) No. 528/2012, Evaluation of the active substance Propiconazole for product-type 7 (film preservatives), evaluating Competent Authority: Finland (January 2015); available online: http://echa.europa.eu/information-on-chemicals/biocidal-active-substances.
[243] European Chemicals Agency (ECHA) Website: Committee for Risk Assessment (RAC), Opinion proposing harmonised classification and labelling at EU level of propiconazole (adopted 9. 12. 2016); available online: https://echa.europa.eu/de/opinions-of-the-committee-for-risk-assessment-on-proposals-for-harmonised-classification-and-labelling?p_p_id=viewsubstances.
[244] European Chemicals Agency (ECHA) Website, Substance information: Diuron (19. 04. 2016); available online: http://echa.europa.eu/in/substance-information/-/substanceinfo/100.005.778.
[245] EFSA Scientific Report (2005) 25, p. 1–58, Conclusion on the peer review of Diuron; available online: http://www.efsa.europa.eu/sites/default/files/scientific_output/files/main_documents/25r.pdf.
[246] Commission Implementing Regulation (EU) 2016/872 of 1 June 2016 concerning the non-renewal of approval of the active substance isoproturon, in accordance with Regulation (EC) No. 1107/2009 of the European Parliament and of the Council concerning the placing of plant protection products on the market, and amending Commission Implementing Regulation (EU) No. 540/2011, Official Journal of the European Union L 145/7 (02.06.2016); available online: http://eur-lex.europa.eu/legal-content/EN/ALL/?uri=CELEX:32016R0872.
[247] ECB – Summary fact sheet PBT working group (PBT list No.122) TC NES subgroup on identification of PBT and vPvB substances (last update: 15.09.2015): available online: http://echa.europa.eu/documents/10162/21973129/pbt_122_cas_330-54-1_en.pdf.
[248] EFSA (European Food Safety Authority), 2015. "Conclusion on the peer review of the pesticide risk assessment of the active substance isoproturon", EFSA Journal 2015; 13(8): p. 4206; available online: http://www.efsa.europa.eu/de/efsajournal/pub/4206.
[249] Submitted CLH proposal: Isoproturon; Dossier submitted by Germany (February 2015); available online: http://echa.europa.eu/registry-of-submitted-harmonised-classification-and-labelling-intentions/-/substance-rev/6113/term.
[250] U.S. National Library of Medicine, National Centre for Biotechnology Information, Compound Summary for CID 13450 (Terbutryn) – 22.04.2016; available online: https://pubchem.ncbi.nlm.nih.gov/compound/Terbutryn.
[251] F. Tettenborn, T. Hillenbrand, E. Menger-Krug, J. Niederste-Hollenberg, S. Gärtner, P. Baur, J. Jäger (2014). "Neue prioritäre/prioritär gefährliche Stoffe der Richtlinie 2013/39/EU des Europäischen Parlaments und des Rates – Stoffdatenblätter", Fraunhofer-Institut für System- und Innovationsforschung (ISI), im Auftrag des Umweltbundesamtes; available online: http://www.isi.fraunhofer.de/isi-wAssets/docs/n/de/publikationen/fkz_3709_67_219_emissionen_anhang_g_bf.pdf.
[252] European Chemicals Agency (ECHA) Website, Classification & Labelling Inventory, Terbutryn (22.04.2016); available online: http://echa.europa.eu/information-on-chemicals/cl-inventory-database/-/discli/details/69021.
[253] U.S. Environmental Protection Agency, Office of Pesticide Programmes, Chemicals Evaluated for Carcinogenic Potential, Annual Cancer Report 2015, available online: http://npic.orst.edu/chemicals_evaluated.pdf.
[254] C. D. Klaassen (Ed.) (2001). "Casarett and Doull's Toxicology. The Basic Science of Poisons", (6th Edition) New York, NY: McGraw-Hill, p. 791.
[255] Y. E. Lin (2010). "Legionella in Water Systems", in: The science and technology of industrial water treatment. CRC Press, Boca Raton, FL 33487-2742, p. 411–424, ISBN 978-1-4200-7145-0.
[256] T. Wyke for MailOnline and Associated Press (2015). "GlaxoSmithKline closes down asthma inhaler factory amid fears of Legionnaires' outbreak which left 850 employees at risk", available online: http://www.dailymail.co.uk/news/article-3194755/Drug-factory-makes-asthma-inhalers-closes-amid-fears-Legionnaires-outbreak-North-Carolina-base.html.
[257] C. J. Nalepa, T. M. Williams (2010). "Biocides: Selection and Application", in: The science and technology of industrial water treatment. CRC Press, Boca Raton, FL 33487-2742, p. 381–410, ISBN 978-1-4200-7145-0.
[258] C. P. Gerba (2000). "Disinfection", in: Environmental Microbiology, Maier, R. M., Pepper, I. P., and Gerba, C. P. (Eds.), pp. 543–556, Academic Press, San Diego, CA.
[259] J. Miller, G. D. Simpson (1999). "Chemical control of Legionella", paper presented at the Association of Water Technologies (AWT) Annual Conference, Palm Springs, CA.

[260] B. R. Kim, J. E. Anderson, S. A. Mueller, W. A. Gaines, A. M. Kendall (2002). "Literature review: Efficacy of various disinfectants against Legionella in water systems", Water Research, 36: p. 4433–4444.

[261] M. L. Ludensky, F. J. Himpler, P. G. Sweeny (1998). "Control of biofilms with cooling water biocides", CORROSION/1998, Paper No. 522, NACE International, Houston, TX.

[262] C. J. Nalepa, J. N. Howarth (2002). "Strategies for effective control of surface-associated microorganisms: A literature perspective", paper IWC 02-01, Engineers' Society of Western Pennsylvania, Pittsburgh, PA.

[263] S. Gartiser, E. Ulrich (2002), im Auftrag des Umweltbundesamtes (Ed.). "Einsatz umweltverträglicher Chemikalien in der Kühlwasserkonditionierung", p. 42; available online: http://www.umweltbundesamt.de/sites/default/files/medien/publikation/long/2218.pdf.

[264] B. Blocken, J. Carmeliet (2002). "Spatial and temporal distribution of driving rain on a low-rise building", Wind and Structures, 5(5): p. 441–462. © TECHNO-Press.

[265] F. Sauer (2012). "Testing the tests", European Coatings Journal, 2: p. 28–32.

[266] F. Sauer (2013). "Standard tests overestimate environmental impacts of biocides", Pitture e Vernici – European Coatings, 5: p. 35–42.

[267] S. Gartiser, M. Burkhardt, R. Groß, M. Calliera (2015). "Efficiency and practicability of risk mitigation measures for biocidal products with focus on disinfectants", Final report FKZ 3711 63 410 on behalf of the German Federal Environmental Agency, UBA-Texte 53/2015; available online: http://www.umweltbundesamt.de/publikationen/reduction-of-environmental-risks-from-the-use-of.

[268] M. Burkhardt (2015). "Measured and Predicted Leaching of Film Preservatives in Coatings", HSR University of Applied Science Rapperswil, Institute of Environmental and Process Engineering (UMTEC), 8640 Rapperswil, Switzerland (Cologne, 22th October 2015).

[269] M. Burkhardt, S. Zuleeg, R. Vonbank, K. Bester, J. Carmeliet, M. Boller, T. Wangler (2012). "Leaching of biocides from façades under natural weather conditions". Environmental Science and Technology 46: p. 5497–5503.

[270] European Coatings Website (2015). "The EU Servowood Project – predicting the service life of exterior wood coatings", Vincentz Network; available online: http://www.european-coatings.com/Raw-materials-technologies/Applications/Industrial-coatings/The-EU-Servowood-Project-predicting-the-service-life-of-exterior-wood-coatings.

[271] U. Schoknecht*, H. Mathies, R. Wegner, (2016). "Biocide leaching during field experiments on treated articles", Environ Sci. Eur. 28: p. 6.

[272] U. Schoknecht, H. Mathies, R. Wegner, S. Uhlig, H. Baldauf, B. Colson (2016). "Emissions of material preservatives into the environment – realistic estimation of environmental risks through the improved characterisation of the leaching of biocides from treated materials used outdoors", Final report FKZ 3711 63 411 on behalf of the German Federal Environmental Agency, UBA-Texte 22/2016; available online: http://www.umweltbundesamt.de/publikationen/emissions-of-material-preservatives-into-the.

[273] U. Schoknecht, A. Töpfer, S. Uhlig, H. Baldauf (2012). "Characterisation of leaching of biocidal active substances of main group 2 'preservatives' from different materials under weathering conditions", Final report FKZ 3708 65 404 on behalf of the German Federal Environmental Agency, UBA-Texte 62/2012; available online: http://www.uba.de/uba-info-medien-e/4389.html.

[274] ESTIMATE (2015). "ESTImation of the environmental impact of construction MATErials", available online: http://www.umtec.ch/uploads/tx_hsrpm/Estimate_Flyer.pdf.

[275] K. Breuer, F. Mayer, C. Scherer, R. Schwerd, K. Sedlbauer (2012). "Wirkstoffauswaschung aus hydrophoben Fassadenbeschichtungen: verkapselte versus unverkapselte Biozidsysteme. Bauphysik 34(1): p. 19–23.

[276] H. Rüdel, B. Knopf (2012). "Vorbereitung eines Monitoring-Konzepts für Biozide in der Umwelt", Report FKZ 360 04 036 on behalf of the German Federal Environmental Agency; available online: http://www.umweltbundesamt.de/sites/default/files/medien/384/dokumente/dokument_uba-gutachten_-_monitoring_biozide.pdf.

[277] M. Kahle, I. Nöh (2009). "Biozide in Gewässern – Eintragspfade und Informationen zur Belastungssituation und deren Auswirkungen", Umweltbundesamt-Texte; available online: http://www.umweltbundesamt.de.

[278] M. Andersson Trojer, L. Nordstierna, M. Nordin, M. Nydén, K. Holmberga (2013). "Encapsulation of actives for sustained release", Phys. Chem. Chem. Phys., 15: p. 17727–17741.

[279] J. Bergek, M. Andersson Trojer, H. Uhr, L. Nordstierna (2016). "Controlled release of a microencapsulated arduous semi-hydro-

phobic active from coatings: Superhydrophilic polyelectrolyte shells as globally rate-determining barriers", Journal of Controlled Release 225: p. 31 – 39.

[280] M. Andersson Trojer, L. Nordstierna, J. Bergek, H. Blanck, K. Holmberg, M. Nydéna (2015). "Use of microcapsules as controlled release devices for coatings", Advances in Colloid and Interface Science 222: p. 18 – 43.

[281] F. Sauer (2013). "Squaring the circle: Full potential of biocides in place needed – sound behaviour in environment", European Coatings Congress, Nuremberg (19 March 2013).

[282] A. Lamprecht, R. Bodmeier (2010). "Microencapsulation", in: Ullmann's Encyclopedia of Industrial Chemistry, Wiley-VCH Verlag GmbH & Co. KGaA, Weinheim.

[283] Submitted CLH proposal: Fludioxonil; Dossier submitted by Denmark (September 2015); available online: http://echa.europa.eu/registry-of-submitted-harmonised-classification-and-labelling-intentions/-/substance-rev/10407/term.

[284] D. W. Bartlett, J. M. Clough, C. R. A. Godfrey, J. R. Godwin, A. A. Hall, S. P. Heaney, S. J. Maund (2001). "Understanding the strobilurin fungicides", Pesticide Outlook 12: p. 143 – 148.

[285] D. W. Bartlett, J. M. Clough, J. R. Godwin, A. A. Hall, M. Hamer, B. Parr-Dobrzanski (2002). "The strobilurin fungicides", Pest Manag. Sci. 58: p. 649 – 662.

[286] J. M. Clough, C. R. A. Godfrey et al (1996). "Azoxystrobin: A Novel Broad-Spectrum Systemic Fungicide", Pesticide Outlook 7: p. 16 – 20.

[287] C. D. S. Tomlin (Ed.) (2009). "The pesticide manual: A World compendium", 15th edition, British Crop Protection Council (BCPC), UK.

[288] Mode of action of azoxystrobin: data available from the UK Fungicide Resistance Action Committee website: http://www.frac.info/.

[289] US EPA, Azoxystrobin, Pesticide Fact Sheet, Washington DC, February 1997; available online: https://www3.epa.gov/pesticides/chem_search/reg_actions/registration/fs_PC-128810_07-Feb-97.pdf.

[290] Commission Implementing Regulation (EU) No. 2014/227 of 24 April 2014 on the non-approval of certain biocidal active substances pursuant to Regulation (EU) No. 528/2012 of the European Parliament and of the Council, Official Journal of the European Union L 124/27 (25.04.2014); available online: http://eur-lex.europa.eu/legal-content/DE/ALL/?uri=CELEX:32014D0227.

[291] I. Wiegand, K. Hilpert, R. E. W. Hancock (2008). "Agar and broth dilution methods to determine the minimal inhibitory concentration (MIC) of antimicrobial substances", Nature Protocols 3: p. 163 – 175.

[292] J. M. Andrews (2001). "Determination of minimum inhibitory concentrations", Journal of Antimicrobial Chemotherapy 48 (Suppl. S1): p. 5 – 16.

[293] P. D. Askew (2006). "Coatings and Coatings Materials", in Springer Handbook of Materials Measurement Methods (Eds. H. Czichos, T. Saito, L. Smith), Springer Science and Business Media.

[294] E. Goldman, L. H. Green (2008). "Practical Handbook of Microbiology", 2nd Edition, CRC Press, Taylor and Francis Group, USA.

[295] C. Collins, J. Grange, P. Lyne, J. Falkinham III (2004). "Collins and Lyne's Microbiological Methods", 8th Edition, Arnold, A member of the Hodder Headline Group, London.

[296] K. Winkowski (2001). "Efficacy of in-can preservatives", European Coatings Journal 1 – 2: p. 87 – 91.

[297] J. Lunenburg-Duindam, W. Lindner (2000). "In-can preservation of emulsion paints", European Coatings Journal 3: p. 66 – 73.

[298] Basic Efficacy Method for Biocidal Active Substances used to Preserve Aqueous-Based Products, IBRG PDG16-007; available online: http://ibrg.org/Methods.aspx.

[299] Wet State Paint Method; IBRG P16-001; available online: http://ibrg.org/Methods.aspx.

[300] American Society for Testing and Materials (ASTM), Method D2574 – 06 (2012). "Standard Test Method for Resistance of Emulsion Paints in the Container to Attack by Microorganisms", available online: http://www.astm.org/Standards/D2574.htm.

[301] EN 16105:2011 ("Paints and varnishes – Laboratory method for determination of release of substances from coatings in intermittent contact with water", available online: https://www.beuth.de/de/norm/din-en-16105/140443204.

[302] NT Build 509 (2005). "Leaching of active substances from preservative-treated timber" – Semi-field testing; available online: http://www.nordtest.info/index.php/methods/item/leaching-of-active-ingredients-from-preservative-treated-timber-semi-field-testing-nt-build-509.html.

[303] Commission Regulation (EC) No. 1451/2007 of 4 December 2007 on the second phase of the 10-year work programme referred to in Article 16(2) of Directive 98/8/EC of the European Parliament and of the Council concerning the placing of biocidal products on the market, Official Journal of the European Union L 325/3 (11.12.2007); available online:

http://eur-lex.europa.eu/legal-content/EN/ALL/?uri=OJ:L:2007:325:TOC.

[304] Commission Implementing Regulation (EU) No. 564/2013 of 18 June 2013 on the fees and charges payable to the European Chemicals Agency pursuant to Regulation (EU) No. 528/2012 of the European Parliament and of the Council concerning the making available on the market and use of biocidal products, Official Journal of the European Union L 167/17 (19.6.2013); available online: http://eur-lex.europa.eu/eli/reg_impl/2013/564/oj.

[305] European Chemicals Agency (ECHA) Website, Information on Chemicals, Biocidal Active Substances (19.05.2016); available online: http://echa.europa.eu/web/guest/information-on-chemicals/biocidal-active-substances.

[306] Regulation (EC) No. 1907/2006 of the European Parliament and of the Council of 18 December 2006 concerning the Registration, Evaluation, Authorisation and Restriction of Chemicals (REACH), establishing a European Chemicals Agency, amending Directive 1999/45/EC and repealing Council Regulation (EEC) No. 793/93 and Commission Regulation (EC) No. 1488/94 as well as Council Directive 76/769/EEC and Commission Directives 91/155/EEC, 93/67/EEC, 93/105/EC and 2000/21/EC; consolidated version (10.04.2014) available online: http://eur-lex.europa.eu/legal-content/EN/TXT/?uri=CELEX%3A02006R1907-20140410.

[307] Council Regulation (EC) No. 440/2008 of 30 May 2008 laying down test methods pursuant to Regulation (EC) No. 1907/2006 of the European Parliament and of the Council on the Registration, Evaluation, Authorisation and Restriction of Chemicals (REACH), Official Journal of the European Union L 142/1; consolidated version (04.03.2016) available online: http://eur-lex.europa.eu/legal-content/EN/TXT/?qid=1464346417279&uri=CELEX:02008R0440-20160304.

[308] European Chemicals Agency (ECHA) Website, News and Events: "Does your biocidal product now fall under the scope of the Biocidal Products Regulation?" (10.05.2016); available online: http://echa.europa.eu/view-article/-/journal_content/title/does-your-biocidal-product-now-fall-under-the-scope-of-the-biocidal-products-regulation-.

[309] European Commission Website, Public Health: "Communication from the Commission to the European Parliament and the Council on endocrine disruptors and the draft Commission acts setting out scientific criteria for their determination in the context of the EU legislation on plant protection products and biocidal products" (15.06.2016); available online: http://ec.europa.eu/health/endocrine_disruptors/policy/index_en.htm.

[310] European Commission Website, Public Health: "Endocrine Disruptors – Next steps" (11.06.2017); available online: https://ec.europa.eu/health/endocrine_disruptors/next_steps_en

[311] European Chemicals Agency (ECHA) Website, Biocidal Products Regulation (11.06.2017); available online: http://echa.europa.eu/regulations/biocidal-products-regulation.

[312] Regulation (EC) No. 1107/2009 of the European Parliament and of the Council of 21 October 2009 concerning the placing of plant protection products on the market and repealing Council Directives 79/117/EEC and 91/414/EEC, consolidated version (30.06.2014); available online: http://eur-lex.europa.eu/legal-content/EN/TXT/?qid=1467088206407&uri=CELEX:02009R1107-20140630.

[313] Regulation (EU) No. 305/2011 of the European Parliament and of the Council of 9 March 2011 laying down harmonised conditions for the marketing of construction products and repealing Council Directive 89/106/EEC, consolidated version (16.06.2014); available online: http://eur-lex.europa.eu/legal-content/EN/TXT/?qid=1467088863550&uri=CELEX:02011R0305-20140616.

[314] European Commission, EU Pesticides Database (11.06.2017), available online: http://ec.europa.eu/food/plant/pesticides/eu-pesticides-database/public/?event=activesubstance.selection&language=EN.

[315] European Commission Website, Growth: "Draft Commission Implementing Decision not approving 2-methyl-1,2-benzisothiazol-3(2H)-one as an active substance for use in biocidal products of product-type 13" (23.05.2017); available online: http://ec.europa.eu/growth/tools-databases/tbt/nview.cfm?p=EU_467_EN.

Author

Dr. Frank Sauer, born 1961, studied chemistry at the University of Cologne, Germany, and there, he finalised his PhD in Organic Chemistry in 1993. After a further training in business management and international marketing, he started in 1995 his professional career as Assistant Plant Manager in a small-sized company operating in the field of lacquers and coatings for the car repair industry. In October 1999, he changed to Borchers GmbH, a subsidiary enterprise of Bayer Group and consecutively Lanxess Group. There, he was responsible for the development of additives for paints and coatings and in 2002 he became Head of the R&D department. Effective July 2006, he joined the Performance Chemicals Segment of Lanxess as Manager Technical Marketing Coatings in the business unit 'Material Protection Products' and changed internally to the Regulatory Affairs Department as Manager Regulatory Affairs in November 2012.

Index

A

B

C

D

E

F

G

H

I

L

M

N

O

P

Q

R

S

T

U

V

W

Y

Z

www.ingramcontent.com/pod-product-compliance
Lightning Source LLC
LaVergne TN
LVHW081519300926
849161LV00006B/1221